The Best-loved Everyday Dishes

最想學會的家常菜

從小菜到主食一次學透透（中英對照）

作者	洪白陽
翻譯	許妍飛
攝影	廖家威
主編	彭文怡
美術編輯	鄭雅惠
校對	連玉瑩
行銷	洪仔青
企畫統籌	李 橘
總編輯	莫少閒
出版者	朱雀文化事業有限公司
地址	北市基隆路二段13-1號3樓
電話	（02）2345-3868
傳真	（02）2345-3828
劃撥帳號	19234566 朱雀文化事業有限公司
e-mail	redbook@ms26.hinet.net
網址	http://redbook.com.tw
總經銷	
ISBN	978-986-6029-08-0
CIP	427.1
初版三刷	2014.02
定價	350元
出版登記	北市業字第1403號

About買書：

●朱雀文化圖書在北中南各書店及誠品、金石堂、何嘉仁等連鎖書店均有販售，如欲買本公司圖書，建議你直接詢問書店店員，如果書店已售完，請撥本公司經銷商北中南區服務專線洽詢。北區（03）358-9000 中區（04）2291-4115 南區（07）349-7445

●●至朱雀文化網站購書（http://redbook.com.tw）可享85折優惠。

●●●至郵局劃撥（戶名：朱雀文化事業有限公司，帳號：19234566），
掛號寄書不加郵資，4本以下無折扣，5～9本95折，10本以上9折優惠。

●●●●周一至周五上班時間，親自至朱雀文化買書可享9折優惠。

The Best-loved Everyday Dishes

最想學會的家常菜

從小菜到主食一次學透透 (中英對照)

Mastering the Culinary Art in One Cookbook, from Appetizers to Main Dishes.

作者 洪白陽 (CC老師)

百吃不膩的家常菜
CC's Delicious Home Cooking

有人問我，希望自己老了以後成為什麼樣的老人呢？

我啊，希望自己會是個優雅、快樂的老人。

多年後，即使已經是髮如雪的老婆婆，

我希望自己仍能在料理教室優雅並愉快地和大家一起享受做菜的樂趣。

寫這本的食譜時，同時我也想，如果每個人都能快樂地做菜，世界一定很美好。

因為，每一個人都開心地為心愛的人做菜，

也就表示每一個人都能吃到心愛的人融入感情所做出來的愛心料理，

這樣子不是大家都很幸福嗎？

想到這裡，我以「百吃不膩的家常菜」為主題設計了一些易學易做的美味料理，

兒時最愛的料理、媽媽的拿手菜、我最受烹飪教室裡學生喜歡的菜……

不時浮現在我的腦海，我一定要在這本書中完整呈現這些好味道。

希望大家都能跟我一樣，樂於做菜給家人和朋友吃，

愛上料理是幸福的開始，而我，想一輩子都沉浸在這份幸福裡。

很感謝瑞康鍋具（瑞康國際有限公司）的總經理Vivia蔡，

她提供了我一個完美的烹飪舞台，我才能夠將如此美味的料理呈現在大家眼前。

在此，我也要感謝我的助手，吳杰修先生，

多虧有他，我才能順利地完成這次食譜的拍攝。

當然，更要謝謝各位讀者，

讓我能夠一直享受製作美味料理、品嘗可口菜餚的幸福。

People asked me what kind of person I would like to be when I grow old. I portrait myself to be a graceful and happy senior. Many years after, even if I become a gray-haired grandma, I still hope I will be able to share the joys of cooking with everyone in my class.

As I was writing this cookbook, I kept thinking how wonderful this world would be if everyone experiences the joys of cooking. If everyone cooks happily for the beloved ones, it will be a blessing for all because they will be able to enjoy a meal full of love made by their beloved ones. Thus, all of us are happy, right?

Thinking of this, I designed some simple and tasty dishes based on the theme of "delicious home cooking", My childhood favorites, Mom's specialties, popular dishes among my students in my cooking class, etc., all of which kept emerging in my mind, and made me long to fully showcase the deliciousness to you in this book. I hope everyone can be like me, love to cook for beloved one and enjoy the food made of love. Falling in love with cooking is the beginning of happiness, and as for me, I would love to be immersed in this kind of happiness forever and ever.

I would like to thank Vivia Tsai, the executive general manager of Rakenhouse (Raken International Co., Ltd.), for giving me such a perfect stage to showcase these delicious dishes to everyone. Furthermore, I would like to thank my assistant, Mr. Jié-siou Wu. Thanks to his help so that the photographs used in this book could be successfully taken.

Contents 目錄

油條鮮蚵

Contents 目錄

Part 3
肉類
Meat Section

怪味雞

湘味蒜辣雞丁

清燉牛肉湯

Part 5
湯類
Soup Section

台南擔仔麵

Vegetables and Eggs Section

Part 1 蔬菜、豆腐、蛋類

如果成天只吃大魚大肉，就算是山珍海味，
不用多久，肯定也會倒胃口，營養更是不均衡，
所以餐桌上絕對少不了陪襯的「綠葉」——蔬菜；
還有營養價值高、可隨意料理的蛋，
更是餐桌上不可缺的明星角色。
除了美味可口，更要兼顧營養均衡喔！

It' won't be long for us to lose appetite
if we keep eating fish and meat all the time, even if they are delicious.
Besides, your nutrition intake will be unbalanced.
Therefore, vegetables are a must "side dish" on our dining tables.
Eggs are also the indispensable star.
They are nutritious and versatile in cooking.
Remember to balance the nutrition while savoring delicious food.

素炒三絲

Stir-Fried Chinese Cabbage, Enoki Mushroom and Carrots

這道菜做法簡單又好吃，是每次餐聚素食朋友必向我點的一道料理。也可更改成自己喜歡的素食材，切成絲一起烹調即可。

This dish is delicious and easy to make. My vegetarian friends always ask me to cook this dish every time we have a feast gathering. You can substitute the preferred shredded vegetables for the original ingredients.

材料

大白菜200克、金針菇200克、胡蘿蔔100克、嫩薑30克

調味料

素蠔油3 1/2大匙、香油1大匙

做法

1. 大白菜的白梗部分切絲；胡蘿蔔和嫩薑都切絲。
2. 鍋燒熱，倒入1 1/2大匙油，先放入嫩薑絲炒香，續入大白菜梗炒一下，再放入胡蘿蔔炒約1分鐘半，倒入金針菇快炒1分鐘，最後倒入素蠔油調味，盛盤後淋上香油即可食用。

Ingredients

200g Chinese cabbage, 200g enoki mushrooms, 100g carrots, 30g tender ginger

Seasonings

3 1/2T vegetarian oyster sauce, 1T sesame oil

Directions

1. Shred the white parts of Chinese cabbage; shred the carrots and tender ginger.
2. Heat the wok and add 1 1/2T of oil. First, stir-fry the shredded tender ginger. Then add the shredded cabbage and stir-fry for a little bit. Add the carrots and stir-fry for one and half minutes. Put in the enoki mushrooms and stir-fry for 1 minute. Last, season with vegetarian oyster sauce. Drizzle with the sesame oil before serving.

CC 烹調秘訣 | Cooking tips

素蠔油的量可依各品牌的鹹度不同而做調整，本書食譜中的調味料量都僅供參考，建議讀者以自己習慣的品牌口味做增減。

The saltiness of vegetarian oyster sauce varies from brand to brand. Add as needed. The measurements of the seasonings in this book are for reference only, which can be adjusted according to your preference.

涼拌海帶絲
Shredded Seaweed Salad

小吃店裡必備的小菜，做法相當簡單，很適合當下酒菜，也可以換成海帶結來做，大家和我一起試試看！

This appetizer is on every snack restaurant's menu. It's easy to make and goes well with beer. You can replace the main ingredient with seaweed knots. Let's give it a try!

材料
海帶絲300克、蒜末2大匙、紅辣椒末11/2大匙

調味料
醬油3大匙、黑醋3大匙、細砂糖2/3大匙、香油11/2大匙

做法
1. 將海帶絲放入滾水中汆燙，瀝乾水分。
2. 將蒜末、紅辣椒末和調味料倒入海帶絲中拌勻。
3. 將拌好的海帶絲放入冰箱中冷藏約15分鐘後再食用。

Ingredients
300g shredded seaweed, 2T minced garlic, 11/2T minced red chillies

Seasonings
3T soy sauce, 3T black vinegar,2/3T granulated sugar, 11/2T sesame oil

Directions
1. Blanch the shredded seaweed in boiling water. Drain.
2. Toss the seaweed with the minced garlic, minced red chillies and the seasonings.
3. Put the dish in the refrigerator and let it cool for 15 minutes before serving.

CC 烹調秘訣 | Cooking tips

1. 冬天做這道菜時，不需放入冰箱冷藏，在室溫下放約15分鐘即可食用。
2. 這道料理已有鹹度，建議不要再加鹽調味。

1. This dish doesn't need to go into refrigerator in winter. Just let it cool in room temperature for 15 minutes before serving.
2. The dish already contains saltiness. You don't need to add any salt.

油香蒜茸地瓜葉

Sweet Potato Leaves in Garlic Sauce

這道料理是早年台灣鄉村常吃得到的菜，若能加入豬油、蒜茸拌食，更能凸顯道地的台灣風味。地瓜葉是我最愛的綠色蔬菜，簡單的汆燙或快炒，就是一盤好菜。

This is a common dish in the early Taiwan country villages. Mix it with lard and minced garlic and you will have an authentic Taiwanese cuisine.

材料
地瓜葉600克、豬油或橄欖油2大匙

蒜茸醬
醬油膏4大匙、蒜末2大匙、細砂糖1/2大匙

做法
1. 將地瓜葉放入滾水中汆燙一下，燙熟後瀝乾水分，拌入豬油或橄欖油。
2. 蒜茸醬的材料拌勻。
3. 將蒜茸醬倒入一深容器中，鋪入地瓜葉，另取一個盤子將地瓜葉倒扣出來即可食用。

Ingredients
600g sweet potato leaves, 2T lard or olive oil

Garlic Sauce
4T soy sauce paste, 2T minced garlic, 1/2T granulated sugar

Directions
1. Blanch the leaves in boiling water until tender. Drain. Mix it with lard or olive oil.
2. Mix the ingredients of garlic sauce well.
3. Add the garlic sauce into a deep bowl. Place the leaves on top of it. Unmold the dish on another plate. Serve.

 CC 烹調秘訣 | Cooking tips

1. 也可參照p.99炸好雞油，再放入紅蔥頭炸至酥脆成為油蔥酥，以油蔥酥拌地瓜葉會更香。
2. 一般葉菜類蔬菜汆燙時，必須保留清脆的口感，所以不能汆燙太久，但地瓜葉要軟一點才會好吃，所以汆燙的時間稍久一點。

1. You can refer to p.99 to deep-fry chicken first. Then use the same pot of oil to deep-fry shallots into deep-fried shallots. Seasoning the sweet potatoes leaves with fried shallots makes this dish tastier.
2. Leaf greens are usually suggested to be blanched al dente to keep the crunchy texture. But sweet potato leaves taste better when they are blanched until tender. Thus the cooking time needs to be longer.

蠔油芥蘭
Chinese Broccolis in Oyster Sauce

水煮芥蘭菜再淋上以蠔油特調的醬汁，再平凡不過的家常菜，卻更能抓住每個家人的心。

Drizzling oyster sauce on the blanched Chinese broccolis is a very simple family-style dish but always wins every family member's heart.

材料

芥蘭菜300克、蘑菇60克、香油1/2大匙、太白粉1大匙、水3大匙

醬汁

蠔油2大匙、高湯4大匙、蕃茄醬1/2大匙、辣豆瓣醬1小匙、白醋1小匙

做法

1. 芥蘭菜放入滾水中汆燙熟，取出瀝乾水分，剪成一段一段後排入盤中。
2. 蘑菇切片；高湯做法參照p.117。
3. 鍋燒熱，倒入1/2大匙油，先放入蘑菇炒幾下，續入醬汁的材料煮滾，以太白粉水勾芡，淋上香油拌勻，最後淋在芥蘭菜上即可食用。

Ingredients

300g Chinese broccolis, 60g mushrooms, 1/2T sesame oil, 1T tapioca starch, 3T water

Sauce

2T oyster sauce, 4T broth, 1/2T ketchup, 1t spicy soybean paste, 1t white vinegar

Methods

1. Blanch the Chinese broccolis in boiling water. Remove and drain. Cut into sections and arrange on the plate.
2. Slice the mushrooms. See p.117 for the broth recipe.
3. Heat the pan. Add 1/2T of oil. First add the mushrooms and stir-fry briefly. Add the sauce and bring it to boil. Thicken with tapioca starch mixture. Drizzle with the sesame oil and mix well. Last, pour onto the broccolis and serve.

CC 烹調秘訣 | Cooking tips

1. 如果使用的是罐頭蘑菇，切片後直接使用即可，但如果買的是新鮮蘑菇，為了避免蘑菇片變黑，可拌入少許檸檬汁。
2. 通常油炸食物的溫度控制在160～180℃，市面上有的食用油像花生油、大豆沙拉油等，燃點在160℃以下，容易變質，葡萄籽油的燃點是250℃，橄欖油約192℃，較適合油炸。

1. When using canned mushrooms, you can cook them right after slicing them. But if you are using fresh mushrooms, you need to stir in a little bit of lemon juice to prevent the color from turning black.
2. The ideal oil temperature for deep-frying is usually between 160°C to 180°C. Because the flash points of peanut oil and soybean oil...etc. are below 160°C, they tends to go bad easily. Grape seed oil and olive oil is more ideal for deep-frying since their flash points are 250°C and 192°C.

黑胡椒漬毛豆

Pickled Edamames with Black Peppercorns

這道菜在我們家有「聊天之寶」之稱！
可以邊聊天、看電視，邊啃毛豆真是一
大享受，另外，也是宴客上菜前不錯的
配角。

This is our favorite fun-time snack at home!
It's such a joy to eat edamames while chatting
and watching TV. Besides, when you throw a
party, this is a nice choice of appetizer before
serving the meal.

材料

毛豆1,200克、八角6粒、紅辣椒5支

調味料

黑胡椒粒3大匙、鹽21/2大匙、香油2大匙

做法

1. 毛豆洗淨後放入滾水中煮熟，撈出瀝乾水分；紅辣椒切斜段。
2. 將毛豆和八角、紅辣椒和調味料拌勻，置於一旁約2小時即可食用。

這樣做更省時

如果使用休閒鍋或双享鍋烹煮毛豆，只要將毛豆放入鍋內，倒入1/2杯水，蓋上鍋蓋加熱，等冒煙後改小火煮約3分鐘即可。

Ingredients

1,200g edamames, 6 cloves anise, 5 red chillies

Seasonings

3T black peppercorns, 21/2T salt, 2T sesame oil

Directions

1. Rinse the edamames. Cook in boiling water until done. Drain. Cut the red chillies in half crosswise.
2. Combine the edamames, anise, red chillies and the seasonings together. Let it stand for 2 hours and serve.

Time-saving method

When using a Hotpan or Durotherm, you only need to put edamames in the pan. Add 1/2cup of water. Put the lid on and turn on the heat. Turn to low heat when the steam begins to hiss out of the pan and cook for 3 minutes.

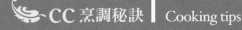

CC 烹調秘訣 | Cooking tips

1. 做好的黑胡椒漬毛豆放入冰箱冷藏，可保存3～7天。
2. 如果使用現磨的黑胡椒粒會更香，建議可至中藥房購買，價格便宜且香氣足。

1. This dish can be stored in the refrigerator for 3 to 7 days.
2. Freshly ground black pepper delivers more kick. You can buy it at Chinese herbal medicine stores at low price and with good quality.

紫茄韭菜

Eggplants with Chinese Chives

我是台灣中部人，家裡的餐桌常會出
現在地口味的料理，這道料理是我從
小吃到大，怎麼都吃不膩的好口味。

I grew up in Central Taiwan. It was
common to have dishes of local flavor on
our dining table. This one is my all-time
favorite. Never get tired of eating it.

材料

茄子600克、韭菜80克、蒜仁3粒、
水1/4杯（1杯約225c.c.）

調味料

醬油1大匙、醬油膏11/2大匙、黑醋
1/2大匙

做法

1. 茄子切長條狀；韭菜切段；蒜仁切
   片。
2. 鍋燒熱，倒入2大匙油，先放入蒜
   片和韭菜頭（白色部分）爆香，續
   入茄子炒一下，倒入水後蓋上鍋蓋
   加熱燜煮。
3. 待茄子煮軟，倒入醬油、醬油膏炒
   勻，放入韭菜，將黑醋先倒在鍋鏟
   裡，再沿著鍋子邊緣淋入拌炒一下
   即可食用。

這樣做更省時

如果使用休閒鍋，水的部分只要1/5杯
即可。

Ingredients
600g eggplants, 80g Chinese chives, 3 cloves of garlic, 1/4 cup of water (1 cup is about 225c.c.)

Seasonings
1T soy sauce, 11/2T soy sauce paste, 1/2T black vinegar

Directions
1. Cut the eggplants into long sections and then each section into slices. Cut the Chinese chives into pieces. Slice the garlic cloves.
2. Heat pan. Add 2T of oil. Stir-fry the sliced garlic and the chive stems (the white part) until fragrant. Add eggplants and stir lightly. Add water, put the lid on and bring to a boil.
3. Once the eggplants are tender, add the soy sauce and soy sauce paste. Stir evenly. Add the Chinese chives. Pour the black vinegar onto a spatula first. Then drizzle it in along the rim of pan. Stir briefly. Serve.

Time-saving method
When cooking in a Hotpan, you only need to add 1/5cup of water.

CC 烹調秘訣 | Cooking tips

1. 喜歡吃辣的人，建議在爆炒蒜片時，一同加入紅辣椒段爆炒即可。
2. 不敢吃韭菜的人，可以改用九層塔，就變成客家風味料理；如果加入辣豆瓣醬，又成了另一
   道好菜。

1. For spicy food lovers: stir-fry garlic slices with red chillies pieces.
2. If you don't like Chinese chives, replace them with basil and it will become a dish with Hakka flavor. You can turn it into another delicious dish by adding spicy soy bean sauce.

烤白菜

Baked Chinese Cabbages

這是以西式料理的做法來烹調白菜，冬天是白菜的季節，一顆顆葉梗新鮮脆嫩的白菜，尤其是製作這道菜的好季節。

This is a Chinese cuisine cooked in western style. Chinese cabbages are in season in winter. It's the ideal time to make this dish with fresh crispy Chinese cabbages.

材料

大白菜600克、麵糊（奶油2大匙、油2大匙、麵粉1/4杯）、無糖奶水1罐、蝦米2大匙、蝦仁100克、鮮干貝100克、起司絲80克、米酒4大匙、高湯1/4杯

調味料

鹽少許、雞湯粉1/2大匙

做法

1. **製作麵糊：** 將奶油、油放入鍋內，以小火加熱，待奶油融化，慢慢加入麵粉攪拌，麵粉加至鍋內呈現類似豆沙般的軟硬度，以小火持續炒至散發出香氣，約10分鐘時間，記得過程中不可用大火。

2. 蝦米洗淨，放入米酒中浸泡約8分鐘後取出；高湯做法參照p.117。

3. 鍋燒熱，倒入2大匙油，先放入蝦米爆香，續入大白菜，淋入米酒，煮至大白菜變軟，加入少許鹽調味，取出大白菜瀝乾水分。

4. 將麵糊、奶水、大白菜的湯汁和高湯倒入鍋中，邊煮邊拌勻至無顆粒狀，這時會呈現出稠狀，加入雞湯粉和30克的起司絲拌勻。

5. 繼續加入大白菜、蝦仁和鮮干貝拌勻，盛入烤皿中，上面鋪上剩餘的起司絲，移入烤箱中，以上下火220℃烤至表面呈焦黃色即可食用。

Ingredients

600g Chinese cabbages, flour batter (2T butter, 2T oil, 1/4cup flour), 1can unsweetened condensed milk, 2T dried shrimp, 100g shrimps, 100g scallops, 80g shredded cheese, 4T rice cooking wine, 1/4cup broth

Seasonings

Salt as needed,1/2T chicken-flavored powder

Directions

1. **Make the flour batter:** Add butter and oil in the pan. Heat over low heat. After the butter is melted, slowly stir in the flour until the firmness of the batter feels like bean paste. Stir-fry on low heat until fragrant, which is about 10 minutes. Remember not to cook on high heat.

2. Rinse the dried shrimps well. Soak in the rice cooking wine for 8 minutes and remove. See p.117 for the broth recipe.

3. Heat the pan and add 2T of oil. First, stir-fry the dried shrimps until fragrant. Then add the Chinese cabbages and the rice cooking wine. Cook until the cabbages are tender. Season with a pinch of salt. Remove the cabbages and drain.

4. Stir in the flour batter, the condensed milk, the cabbage stock and the broth until there is no crumb. When the batter is thickened, stir in the chicken-flavor powder and 30g of shredded cheese.

5. Add the Chinese cabbages, shrimps and scallops and mix well. Pour it on a bakeware. Sprinkle on the remaining shredded cheese and put it in the oven. Bake at 220°C until golden brown and serve.

CC 烹調秘訣 | Cooking tips

1. 麵糊可以多炒一些，放在陰涼處，使用時不要碰到水。
2. 要拌勻麵糊、奶水和高湯時比較難攪拌，建議放入果汁機或食物調理機攪拌比較輕鬆。

1. You can prepare more flour batter at one time and store it in a cool place. Keep the batter away from moisture.
2. It takes some efforts to mix the flour batter, condensed milk and broth evenly. Using a blender or a food processor to mix the ingredients can help you save the sweat.

牛蒡天婦羅

Burdock Root Tempura

雖然我本身不太愛吃牛蒡，但這道牛蒡天婦羅卻是我很喜愛的小食。可以搭配甜辣醬或蕃茄醬食用，當正餐或零食都OK！

I don't like burdock roots, but this burdock root tempura is one of my favorite side dish. It can be served with sweet chili sauce or ketchup. Good for snacks and meals.

材料

牛蒡1支、魚漿600克、味噌1大匙、雞蛋1個、玉米粉3大匙、麵粉1/4杯、白醋3大匙

做法

1. 牛蒡用刀背削除外皮後切絲，放入醋水（1,200c.c.水＋3大匙白醋）中浸泡約6分鐘，防止變成褐色（使偏白色），取出以水清洗，瀝乾水分，撒上麵粉。
2. 將魚漿、味噌和蛋液拌勻，放入玉米粉拌勻，再加入牛蒡絲拌勻。
3. 將做法2.整型成扁圓形，放入油溫160℃的油鍋中炸，炸至呈金黃色後撈出，瀝乾油分，盛盤即可食用。

Ingredients

1 burdock root, 600g fish paste, 1T miso, 1 egg, 3T tapioca starch, 1/4cup flour, 3T white vinegar

Directions

1. Use the back of a kitchen knife to peel the burdock root and slice it. Soak it in the vinegar water (1,200 c.c. water + 3T white vinegar) for 6 minutes. This method can prevent the burdock root from turning brown (and remain the whitish color). Take out the burdock root, rinse well and drain. Sprinkle on flour.
2. Mix the fish paste, miso and whisked egg well. Add tapioca starch and mix well. Add shredded great burdock and mix well.
3. Shape the mixture of **2.** into a flat circle. Deep-fry in hot oil at 160°C until golden brown. Drain and serve.

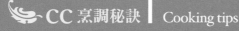

CC 烹調秘訣 | Cooking tips

1. 油炸任何食物時，需等所有的食物都取出後才能熄火，以免食物含過多的油。
2. 牛蒡泡醋水可預防因氧化而變黑，但記得白醋不需放得太多。

1. While deep-frying, never turn off the heat before removing all the food from the fryer. Otherwise, the food will absorb excess oil.
2. Soaking burdock roots in the vinegar water can prevent them from turning brown due to oxidation.

上海南瓜羹

Shanghai-Style Pumpkin Soup

這是一道上海菜，羹湯滑嫩、入口即化，是一道很受長輩歡迎的料理。如果有年長的朋友來作客，相信這道料理會讓賓主盡歡。

This is a Shanghai-style dish. The soup has a silky texture and is easy to swallow. It's very popular among elders. If you have senior guests at home, this will surely be a successful entertaining dish.

材料

南瓜600克、嫩豆腐1塊、生雞胸肉200克、蝦仁150克、高湯1,200c.c.、香菜適量

調味料

鹽適量

做法

1. 嫩豆腐切丁；雞胸肉切末；蝦仁挑除腸泥後切丁；高湯做法參照CC烹調秘訣2.或p.117。
2. 南瓜削除外皮，挖掉籽後切薄片，與高湯一起放入鍋中煮爛。
3. 接著放入豆腐煮約3分鐘，再放入雞胸肉、蝦仁拌勻，煮熟，最後以鹽調味。
4. 將南瓜羹盛入湯碗，撒上香菜即可食用。

Ingredients

600g pumpkin, 1 silken tofu, 200g raw chicken breast, 150g shrimps, 1,200c.c. broth, coriander as needed

Seasonings

Salt as needed

Directions

1. Dice the silken tofu. Mince the chicken breast. Remove and discard the shrimps' sand veins. Dice the shrimps. Please see the second cooking tip or p.117 for the broth recipe.
2. Peel the pumpkin. Remove the seeds. Thinly slice the meat and cook with the broth until mushy.
3. Add the tofu and cook for 3 minutes, then the chicken breast and shrimps and stir evenly. Cook thoroughly. Season with salt.
4. Ladle the soup into a bowl. Sprinkle with coriander and serve.

 CC烹調秘訣 Cooking tips

1. 這道料理在過年製作年菜時，只要多放入蟹肉和鮮干貝，立刻變成黃金一品羹，很受喜愛喔！
2. 我在烹調時，會將剝下來的蝦殼和雞胸肉的雞骨頭放入鍋內，倒入1,500c.c.的水和1/4個洋蔥，熬煮成湯，再以過濾出的高湯來做這道料理。

1. With crab meats and fresh scallops added, this soup will turn into a very popular dish called "Golden Seafood Soup" which can be served in Chinese New Year Feast.
2. When making this dish, I will put the leftover shrimp shells and the chicken bones in a pot. Add 1,500c.c. of water and 1/4 of onion. Bring to a boil and immediately reduce heat to bring the broth to barely a simmer until it's done. Strain the broth and use it for this dish.

豆瓣燒苦瓜

Braised Bitter Squash
in Soybean Paste

這是一道很下飯的家常菜，以豆瓣
醬煮透苦瓜，連不愛吃苦瓜的朋友
也能輕易愛上苦瓜。

This homestyle dish goes well
with rice. By braising bitter squash
thoroughly in soybean paste, even the
non-bitter squashes eaters will easily
fall in love with it.

材料

苦瓜600克、五花肉150克、蒜仁3粒、蔥1支、辣豆瓣醬1 1/2大匙、米酒1大匙、水或高湯1/3杯

調味料

醬油1大匙、醬油膏1大匙、細砂糖1/4小匙

做法

1. 苦瓜洗淨後剖開,去籽後切成1.2～1.5公分的寬條;五花肉切片;高湯做法參照p.117。
2. 鍋燒熱,倒入1 1/2大匙油,放入五花肉爆香至呈金黃色,續入蒜片、蔥段爆香,再加入辣豆瓣醬炒一下,淋入米酒炒勻。
3. 放入苦瓜、水(或高湯)和調味料,蓋上鍋蓋加熱,燜燒至苦瓜變軟即可食用。

這樣做更省時

如果使用休閒鍋,蓋上鍋蓋加熱,只要煮約6分鐘,再移入外鍋燜約4分鐘即可。

Ingredients

600g bitter squash, 150g pork belly, 3 garlic cloves, 1 green onion, 1 1/2T spicy soybean paste, 1T rice cooking wine, 1/3cup of water or broth

Seasonings

1T soy sauce, 1T soy sauce paste, 1/4t granulated sugar

Directions

1. Rinse the bitter squash and split in half. Remove seeds. Cut it into 1.2 to 1.5 cm-wide board strips. Slice the pork belly. See p.117 for the broth recipe.
2. Heat the pan and add 1 1/2T of oil. Stir-fry the sliced pork belly until golden brown, then add the sliced garlic and green onion sections and stir-fry until fragrant. Add the spicy soybean paste and stir briefly. Drizzle the rice cooking wine and mix well.
3. Add the bitter squash, water (or broth) and seasoning. Put the lid on and turn the heat higher. Braise until the bitter squash is tender. Serve.

Time-saving method

If you use a Hotpan, put the lid on and turn on the heat. After cooking for 6 minutes, remove to an outer pot and set aside for 4 minutes (with the lid on) before serving.

CC 烹調秘訣 | Cooking tips

1. 不吃辣的人,改成一般的豆瓣醬即可。
2. 為了健康,我烹調中式料理不用味精,在調味上改用少許(幾顆粒)的糖,可讓料理更鮮美,而且吃不出糖的甜味喔!

1. For non-spicy food eaters, just substitute the regular soybean paste for the spicy one.
2. For the health concerns, no MSG is added in my Chinese cuisine. I replace it with a little bit (just a few drops) of sugar to make the dish taste more delicious. You won't be able to taste the sweetness of sugar.

鹹蛋香酥杏鮑菇

Crispy King Oyster Mushrooms with Salted Eggs

這道菜的主角是炸得香酥的杏鮑菇與鹹蛋，加上特殊香氣的九層塔為配角，樸實外表下卻能散發出食材的美味。

The highlight of this dish is the crispy king oysters and salted eggs. Accompanied with the special fragrance of the basil, this simple dish can truly showcase the tastiness of the ingredients.

材料

杏鮑菇300克、九層塔1把（手抓一把握住拳頭的量）、蒜仁3粒、紅辣椒2支、鹹蛋2個、香油1/2小匙、薑片5片、雞蛋2個、地瓜粉1/2杯、麵粉適量

調味料

鹽、白胡椒粉、細砂糖各少許

做法

1. 杏鮑菇切滾刀塊；蒜仁切片；紅辣椒切斜段；鹹蛋剝殼後切小丁。

2. 將杏鮑菇依序「沾裹麵粉→蛋液→地瓜粉」，放入油溫160℃的油鍋中炸至呈金黃色，取出瀝乾油分。

3. 鍋燒熱，倒入11/2大匙油，先放入蒜片、薑片爆香，續入紅辣椒炒一下，再放入鹹蛋炒一下，倒入杏鮑菇，倒入調味料拌勻後熄火。

4. 撒上九層塔，淋上香油即可食用。

Ingredients

300g king oyster mushrooms, 1 bunch of basil leaves (about a handful of), 3 cloves of garlic, 2 red chillies, 2 salted eggs, 1/2t sesame oil, 5 slices of ginger, 2 eggs, 1/2cup of sweet potato flour, flour as needed

Seasonings

Salt, white pepper and granulated sugar as needed

Directions

1. Roll cut the king oyster mushrooms. Slice the garlic cloves. Cut red chillies into sections. Peel and dice the Salted eggs.

2. Coat the mushrooms first with flour, then a layer of whisked egg and last sweet potato flour. Deep-fry the mushrooms in hot oil at 160°C until golden brown. Drain.

3. Heat the pan. Add 11/2T of oil. Stir-fry sliced garlic and ginger until fragrant. Add red chillies and stir-fry briefly. Then add salted eggs and stir-fry for a little bit. Add the king oyster mushrooms and the seasonings. Cook until evenly mixed and turn off the heat.

4. Sprinkle on basil. Drizzle with the sesame oil. Remove and serve.

CC 烹調秘訣 | Cooking tips

1. 油炸食物時，需將所有食物都取出後才可以熄火，以免食物吸入過多的油。

2. 油溫160℃的判斷方法，可試著將一塊地瓜粉漿放入加熱的油鍋中，如果立刻冒出很多泡泡，代表油溫已經達到160℃，如果是慢慢起泡，大約是120℃。

3. 砂糖在這道料理中僅是取代味精，所以不用加太多，只要加入幾顆或少許即可。

1. While deep-frying, turn off the heat after all the food is removed from the fryer. Otherwise, the food will absorb excess oil.

2. How to tell the oil temperature reaches 160°C: drop a piece of sweet potato flour batter into the preheated oil. If the oil bubbles vigorously, it means the temperature reaches 160°C. If the bubbles pop up slowly, the temperature is about 120°C.

3. Only a few drops or a little bit of sugar is needed in this dish. The purpose of adding sugar is to replace the monosodium glutamate, also known as MSG, thus you don't need to add too much of it.

客家風味燒豆腐

Hakka-Style Braised Tofu

客家菜最大的特色在於偏鹹、重口味和香濃，是很下飯的配菜。尤其缺乏食慾的人，來點客家菜一定讓你胃口大開。

The characteristics of Hakka cuisine are more salty, heavy-flavored and with a pleasant, strong fragrance. For those who don't feel like eating, this dish will definitely increase your appetite.

材料

板豆腐2塊、五花肉100克、蒜苗3
支、紅辣椒2支、豆豉11/2大匙、高
湯或水1/4杯（1杯約225c.c.）、米酒
11/2大匙、香油1大匙、太白粉1大匙

調味料

醬油1大匙、醬油膏1大匙、白胡椒粉
少許

做法

1. 板豆腐切長條狀；五花肉切片；1
支蒜苗切絲；2支蒜苗和紅辣椒切
段；豆豉放入米酒中浸泡約8分鐘
後取出；高湯做法參照p.117。

2. 鍋燒熱，倒入2大匙油，放入板豆
腐煎至兩面都呈金黃色，取出。

3. 放入五花肉，煎至呈金黃色，立刻
加入豆豉爆香，放入蒜苗段和紅辣
椒段、調味料和高湯煮滾，再放入
板豆腐煮約10分鐘。

4. 加入太白粉勾薄芡，淋上香油後盛
盤，撒上蒜苗絲即可食用。

Ingredients

2 squares firm tofu, 100g pork belly, 3 leeks, 2 red chillies,
11/2T fermented soybeans, 1/4cup of broth or water (1 cup
is about 225c.c.), 11/2T rice cooking wine, 1T sesame oil, 1T
tapioca starch

Seasonings

1T soy sauce, 1T soy sauce paste, white pepper as needed

Directions

1. Cut the firm tofu into long sections. Slice the pork belly.
Shred one of the leeks. Cut the other two leeks and red
chillies into sections. Soak the fermented soybeans in the
rice cooking wine for 8 minutes. Drain. Please see p.117
for the broth recipe.

2. Heat the pan and add 2T of oil. Sauté the tofu until two
sides are golden brown. Remove.

3. Sauté the pork belly until golden brown. Immediately add
the fermented soybeans and stir-fry until fragrant. Add the
leeks, red chillies, seasonings and broth. Bring to a boil,
add tofu and cook for 10 minutes.

4. Thicken with tapioca starch mixture. Drizzle with the
sesame oil and scoop the food onto a serving plate.
Sprinkle on the shredded leek and serve.

CC 烹調秘訣 | Cooking tips

1. 將豆豉以米酒浸泡，可使豆豉味道更香醇。如果買到的豆豉比較鹹，醬油要酌量加入。
2. 板豆腐除了用煎的，也可以放入烤箱，以上下火200℃烤至呈金黃色，更帶香氣。

1. Soaking fermented soybeans in rice cooking wine will intensify its fragrance. If the fermented
soybeans that you bought taste too salty, you will need to reduce the amount of soy sauce.
2. In addition to sauté, you can also bake the tofu in the oven at 200°C until golden brown to bring
out its fragrance.

麻婆豆腐

Ma-Po Tofu

記得我在高中一年級做這道有名的四川菜時，很受附近眷村媽媽的讚賞，可以說是最拿手的中式料理之一。

I remember when I first made this famous Sichuan dish in 10th grade, it won generous praise from the neighbors in Armed Forces Military Department Quarters. It is probably one of my signature Chinese cuisines.

材料

豆腐2塊、絞肉80克、蔥末1大匙、蒜末1大匙、薑末1大匙、紅辣椒末1大匙、辣豆瓣醬1 1/2大匙、甜麵醬1大匙、米酒1大匙、高湯1杯、太白粉1 1/2大匙、白醋1/2大匙、香油1大匙、花椒粉少許、蔥花1大匙

調味料

醬油1大匙、醬油膏1 1/2大匙

做法

1. 豆腐放入加些許鹽的滾水中稍微汆燙一下，撈出切塊狀；高湯做法參照p.117。
2. 鍋燒熱，倒入2大匙油，先放入絞肉以中小火炒散，續入蔥末、蒜末、薑末和紅辣椒末爆香，倒入辣豆瓣醬、甜麵醬炒勻，淋入米酒拌炒均勻，再倒入高湯煮滾，放入豆腐、調味料，以小火煮約12分鐘，加入太白粉勾芡，淋入鍋邊醋（白醋），撒入香油。
3. 起鍋盛入盤中，撒入花椒粉和蔥花即可食用。

Ingredients

2 squares tofu, 80g ground pork, 1T minced green onions, 1T minced garlic, 1T minced ginger, 1T minced red chillies, 1 1/2T spicy soybean paste, 1T sweet bean paste, 1T rice cooking wine, 1 cup broth, 1 1/2T tapioca starch , 1/2T white vinegar, 1T sesame oil, Sichuan peppercorn powder as needed, 1T chopped green onions

Seasonings

1T soy sauce, 1 1/2T soy sauce paste

Directions

1. Quickly blanch the tofu in slightly salty and boiling water. Remove and cut it in cubes. See p.117 for the broth recipe.
2. Heat the pan. Add 2T of oil. First, stir-fry the ground pork on medium-low heat. Then add the minced green onions, garlic, ginger and red chillies and stir-fry until fragrant. Add the spicy soybean paste and sweet bean paste and mix evenly. Drizzle with the rice cooking wine and mix evenly. Next add the broth and bring to a boil. Add tofu, seasonings and cook 12 minutes in low-heat. Add tapioca starch to thicken. Drizzle with the white vinegar along the rim of pan. Drizzle with the sesame oil.
3. Remove from heat. Sprinkle with the Sichuan peppercorn powder and the chopped green onions.

CC 烹調秘訣 | Cooking tips

1. 豆腐先以鹽水煮過比較不易破碎。烹調時，需燜10分鐘以上才會入味。
2. 做法2.中的鍋邊醋，是指將白醋先倒在鍋鏟裡，再沿著鍋子邊緣淋1/3圈，這個動作可增加料理的香氣。
3. 放入豆腐翻動時，需用鍋鏟的背面，以鏟圓的方式將豆腐推往鍋邊，才不會將豆腐弄碎。

1. Blanching tofu in salty water first can help tofu hold its shape. When cooking tofu, make sure it is simmered for more than 10 minutes so that the flavor will be able to penetrate.
2. How to drizzle the vinegar along the rim of pan: Pour white vinegar on a spatula first, then drizzle along the pan for 1/3 circle to bring out the flavor of this dish.
3. Use the back of a spatula to stir tofu. Stirring tofu from center to edge of the pan can help tofu keep its shape.

芹香菜脯蛋

Radish Omelet with Minced Celery

這道菜脯蛋料理和一般的菜脯蛋略不同，變得更好吃！秘訣在於多加了芹菜末和鮮奶油這2個法寶，趕快來試試！

This dish is different than other radish omelets because it's tastier! The secret is the minced celery and whipped cream. Let's give it a try!

材料

珍珠菜脯200克、雞蛋5個、芹菜末2大匙、蔥花2 1/2大匙、鮮奶油2大匙

做法

1. 珍珠菜脯洗淨，瀝乾水分；雞蛋打散。
2. 鍋燒熱，倒入1大匙油，放入菜脯炒約3分鐘至散發出香氣，取出。
3. 將蛋液放入容器中，加入鮮奶油先攪打均勻，再放入芹菜末、蔥花和做法2.拌勻。
4. 鍋燒熱，倒入3大匙油，倒入做法3.兩面煎熟，盛盤即可食用。

Ingredients

200g pearl radishes, 5 eggs, 2T minced celery, 2 1/2T chopped green onions, 2T whipped cream

Directions

1. Rinse the radishes well and drain. Whisk the eggs.
2. Heat the pan. Add 1T oil. Add the radishes and stir-fry 3 minutes until fragrant. Remove.
3. Pour the whisked egg in a bowl. Stir in the whipped cream. Fold in the minced celery, chopped green onions and the radishes in direction **2**. Whisk evenly.
4. Heat the pan. Add 3T of oil. Add the radish mixture in direction **3**. Fry the bottom and the top of it until done. Remove and serve.

 CC 烹調秘訣 | Cooking tips

1. 菜脯要先炒過才會香，加入蛋液煎成菜脯蛋才好吃。此外，一般炒蛋都會用較多油。
2. 珍珠菜脯是指將菜脯切成小丁狀販售，使用上比較方便，但如果買到的是長條狀，可自己切成丁狀即可。

1. Stir-frying radishes in advance can bring out its flavor and make omelets taste better after combining with whisked egg.
2. Pearl radishes refer to the radishes that have been precut into dices. You can find them in stores. It's easier to use. If the radishes that you buy are in the shape of a strip, just dice them yourself.

香香蛋

Aromatic Eggs

這道蛋料理是我在高中時代學會
的，最特別的地方在於一股特殊的
香氣，令人吃過難忘！做法很簡
單，推薦給烹飪新手嘗試。

I learned how to make this egg dish
when I was a high school student.
What's so special about it is the special
fragrance. This is the dish that would
be hard to forget. The directions are
very simple. Good for beginners.

材料

雞蛋6個、蝦米2大匙、芹菜末2大匙、蔥末4大匙、香菜末2大匙、米酒2大匙、油蔥酥1 1/2大匙

調味料

醬油2 1/2大匙

做法

1. 蝦米洗淨，放入米酒中浸泡約8分鐘，取出切碎；油蔥酥做法參照p.99。
2. 鍋燒熱，放入蝦米，以小火炒至散發出香氣。
3. 將除了米酒以外的所有材料倒入容器中，倒入醬油拌勻。
4. 鍋燒熱，倒入3大匙油，倒入做法3.煎熟，盛盤即可食用。也可煎成蛋卷，再切成壽司狀。

Ingredients
6 eggs, 2T dried shrimps, 2T minced celery, 4T minced green onions, 2T minced coriander, 2T rice cooking wine, 1 1/2T deep-fried shallots

Seasonings
2 1/2T soy sauce

Directions
1. Rinse the dried shrimps and soak in the rice cooking wine for 8 minutes. Remove and chop. See p.99 for the deep-fried shallots recipe.
2. Heat the pan. Add the dried shrimps and stir-fry on low heat until fragrant.
3. Put all the ingredients except the rice cooking wine into a bowl. Toss with soy sauce well.
4. Heat the pan. Add 3T of oil. Sauté the egg mixture in direction 3. until done. Remove and serve. You can also make it into the shape of omelet and slice it into the shape of sushi.

CC 烹調秘訣 | Cooking tips

1. 蝦米以米酒先浸泡過，除了可以軟化蝦米，還可去除腥味以增加香氣。
2. 如果有鮮奶或鮮奶油，或者搭配咖啡的奶油球，可加入2大匙在做法3.內一起拌勻，口感更軟綿且增加香氣。
3. 煎香香蛋不可用大火，尤其是要煎成蛋卷，易導致外層焦黃而裡面沒熟。

1. Soaking dried shrimp in rice cooking wine in advance can not only soften dried shrimps but also remove the fishy smell to increase its aroma.
2. If you happen to have milk, whipped cream, or liquid coffee creamer, stir in 2T in direction 3. It will contribute to a creamier texture and rich aroma.
3. Never cook aromatic eggs on high heat, especially making it into omelet. You will easily burn the outside while the inside remains undone.

Seafood Section

Part 2 海鮮類

上館子吃飯的時候，總少不了一道海鮮料理。
或清蒸或紅燒，不論什麼烹調方式，
那一股鮮美總教人忍不住一口接一口。
那細緻的肉質、唇齒留香的美味，
讓海鮮成為不輸給肉料理的另一個餐桌亮點！

Seafood is a dish that we definitely will order when dine out.
Either steamed or braised, the fresh and deliciousness of seafood
always makes you eat one bite after another.
Its delicate texture and the wonderful flavor
that lingers in your mouth make it another highlight on dining tables
both at home and at restaurants.

五味花枝
Calamari in Five Flavor Sauce

酸甜辣的五味醬適合搭配花枝、章魚、鮮蚵、生蠔和淡菜等海鮮類食用，但淋在煮熟的雞肉、茄子、豬肉和炸得酥脆的豆腐也很可口。

Five-flavor sauce tastes sweet, sour and spicy. It not only goes well with seafood like calamari, squids, oysters and mussels but also makes cooked chicken, eggplants, pork and crispy deep-fried tofu taste delicious.

材料

花枝或軟絲600克、嫩薑1小塊

五味醬

蔥末1大匙、嫩薑末1大匙、紅辣椒末1大匙、蒜末1大匙、香菜末1大匙、香油1大匙、醬油膏3 1/2大匙、糖1/2大匙、黑醋1/2大匙、蕃茄醬2大匙

做法

1. 花枝切塊狀，放入滾水中煮熟後取出，瀝乾水分，盛盤。
2. 嫩薑切細絲。
3. 將五味醬拌勻，淋在花枝上，盤邊放上嫩薑絲即可。

Ingredients

600g calamari or squid, 1 small piece of tender ginger

Five-flavor sauce

1T minced green onions, 1T minced tender ginger, 1T minced red chillies, 1T minced garlic, 1T minced coriander, 1T sesame oil, 3 1/2T soy sauce paste, 1/2T sugar, 1/2T black vinegar, 2T ketchup

Directions

1. Cut the calamari into bite-size pieces. Blanch in boiling water until cooked. Remove, drain and place on a serving plate.
2. Shred the tender ginger.
3. Mix the five-flavor sauce well. Pour on the calamari. Garnish with some shredded ginger on the side.

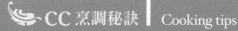

CC 烹調秘訣 | Cooking tips

1. 可以用蒜苗取代蔥。
2. 五味雞腿DIY：將雞腿撒上少許鹽和白胡椒粉，置於一旁約15分鐘。鍋燒熱，倒入1大匙油，放入雞腿煎熟，取出切塊狀後盛盤，淋上五味醬即可。

1. You can substitute green onions with leek.
2. **How to make sautéed chicken in five-flavor sauce:** Rub a pinch of salt and white pepper on chicken quarter legs then set aside for 15 minutes. Heat the pan. Add 1T oil and sautéed the chicken until cooked thoroughly. Cut into pieces and place on a serving plate. Pour on the five-flavor sauce and it's ready.

蒜苗炒鯊魚
Stir-Fried Shark Meat with Leeks

在鯊魚的多種烹調方式中，快炒是最能呈現出食材原味，以及縮短烹煮時間的好方法。

There are many ways of cooking shark meat. However, stir-fry is the best way to showcase the original flavor of ingredients and shorten the cooking time.

材料

鯊魚300克、蒜苗2支、芹菜4支、紅辣椒2支、嫩薑片2片、米酒1大匙

調味料

鹽適量、白胡椒粉少許

做法

1. 鯊魚切約3×3公分的塊狀；蒜苗、芹菜和紅辣椒都切段。
2. 鍋中倒入1,000c.c.水煮滾，先放入薑片稍微煮一下，再放入鯊魚汆燙，取出。
3. 鍋燒熱，倒入2大匙油，先放入蒜白爆香，續入紅辣椒和鯊魚，淋入米酒快炒幾下，倒入調味料，最後放入蒜青、芹菜稍微翻炒即可食用。

Ingredients

300g shark meat, 2 stalks of leeks, 4 stalks of celery, 2 red chillies, 2 sliced tender ginger, 1T rice cooking wine

Seasonings

Salt as needed, white pepper to taste

Directions

1. Cut the shark meat into 3x3cm squares. Cut the leek, celery and red chillies into sections.
2. Boil 1,000c.c. of water in a pot. First, add the sliced ginger and cook for a little bit. Then, blanch the shark meat. Remove.
3. Heat the pan. Add 2T oil. Stir-fry the white parts of leeks first until fragrant. Add the red chillies and shark meat. Pour on the rice cooking wine and stir-fry briefly. Add the seasonings. Last, add the green parts of leeks and celery. Stir-fry slightly and serve.

 CC 烹調秘訣 | Cooking tips

蒜青（蒜苗綠色的部分）不可炒太久，會變得很粗糙而口感不佳。
If the green part of leek is overcooked, the texture will turn rough and make the leek taste bad.

油條鮮蚵

Oyster and Chinese Crullers

台灣的蚵仔，味鮮肉飽滿一點都不
輸給國外進口的生蠔。新鮮蚵仔可
以煮湯、煮粥、生炒來烹調，料理
方式多。這一道油條鮮蚵，淋上特
調醬汁，最下飯。

The juicy and meaty domestic oysters
taste as good as imported ones. There
are many ways to cook fresh oysters,
such as soups, congee and sauté dishes.
After pouring on the special sauce, this
oyster with Chinese crullers dish goes
very well with rice.

材料

蚵仔300克、油條1根、嫩薑適量、香菜適量、蔥花2大匙、麵粉4大匙、酒2大匙

淋醬

醬油膏2大匙、素蠔油2大匙、蒜末2大匙、嫩薑末1大匙、細砂糖1/2大匙、檸檬汁1/2大匙、香油1/2大匙

做法

1. 蚵仔用麵粉和酒稍微拌一下，不可太用力，再放在水龍頭下清洗。注意水不能太大，以細小水量清洗，再瀝乾水分。
2. 將蚵仔放入滾水中汆燙一下（不要太熟），瀝乾水分。
3. 油條切約2公分長；嫩薑切絲；淋醬的材料拌勻。
4. 將油條鋪在盤底，放上蚵仔，淋上淋醬，撒上香菜、蔥花，盤邊放些嫩薑絲即可食用。

Ingredients

300g oysters, 1 Chinese cruller, tender ginger as needed, coriander as needed, 2T finely chopped green onions, 4T flour, 2T wine

Sauce

2T soy sauce paste, 2T vegetarian oyster sauce, 2T minced garlic, 1T minced tender ginger, 1/2T granulated sugar, 1/2T lemon juice, 1/2T sesame oil

Directions

1. Slightly mix the oysters with flour and wine. Don't press it too hard. Rinse in running tap water. Don't turn on too much water. Wash in a small stream of water. Drain.
2. Blanch the oysters in boiling water quickly (and gently). Drain.
3. Cut the Chinese cruller in 2cm sections. Shred the tender ginger. Combine the sauce.
4. Place the Chinese cruller sections on a plate. Add the oysters. Pour on the sauce. Sprinkle with coriander and finely chopped green onions. Garnish with some shredded tender ginger on the side of plate. Serve.

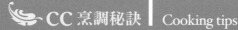 **CC 烹調秘訣** | Cooking tips

買回來的油條如果變軟口感不脆，或者太油，可將油條放入烤箱中，以上下火180℃烤6～8分鐘。

If the Chinese crullers that you buy have lost its crunchiness or become too greasy, just put it in the oven and bake at 180°C for 6 to 8 minutes.

油燜冬瓜
鮮干貝

Oil-Simmered White Gourd
with Fresh Scallops

口感綿密的冬瓜雖然本身沒什麼味
道，但加入鮮干貝一起烹調，冬瓜
變得不一樣囉！鮮美的湯汁更是不
能浪費喔！

White gourd has delicate texture and
doesn't have much flavor itself. However,
it tastes so flavorful after cooking with
fresh scallops. Remember not to waste
the delicious stock!

材料

冬瓜900克、嫩薑片5片、小顆鮮干貝120克、蠔油3大匙、水或高湯1/2杯

做法

1. 冬瓜削除外皮後去籽，切成5X5公分的大塊狀；高湯做法參照p.117。
2. 鍋燒熱，倒入2大匙油，先放入薑片爆香，續入冬瓜、蠔油和高湯，煮滾後蓋上鍋蓋加熱，轉小火燜煮至冬瓜熟透。
3. 最後加入鮮干貝，煮至鮮干貝熟了即可食用。

這樣做更省時

如果使用休閒鍋或双享鍋，只要加入1/3杯的高湯或水，就能更縮短烹調時間。

Ingredients

900g white gourd, 5 slices tender ginger, 120g small fresh scallops, 3T oyster sauce, 1/2cup water or broth

Directions

1. Peel the white gourd, remove the seed then cut into 5x5 cm pieces. See p.117 for the broth recipe.
2. Heat the pan. Add 2T oil. Stir-fry sliced ginger first until fragrant. Then add white gourd, oyster sauce and broth. Bring to a boil, put the lid on then reduce the heat to low. Simmer until winter gourd is cooked thoroughly.
3. Add the fresh scallops. Cook until scallops are cooked thoroughly and serve.

Time-saving method

When using a Hotpan or Durotherm, you just need to add 1/3 cup of broth or water to reduce the cooking time.

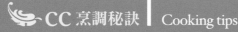

CC 烹調秘訣 | Cooking tips

1. 素食者也想品嘗嗎？放心，這道料理即使不放干貝也很美味，再以素蠔油取代蠔油就可以囉！
2. 冬瓜因為性寒，所以加入薑爆香以去寒。

1. Vegetarian version: this dish still tastes delicious even without scallops. Just substitute vegetarian oyster sauce over oyster sauce.
2. Because winter gourds are cooling foods, we need to add stir-fried ginger to remove excessive "cold" energy from the body.

潮州風味虯目魚

Chaozhou Style Steamed Milkfish

除了香煎虱目魚淋上檸檬汁，或者將煎虱目魚和嫩薑片、醬油和水煮以外，潮洲風味的虱目魚也是很棒的虱目魚吃法！

In addition to grilling milkfish with lemon juice or braising milkfish with tender ginger, soy sauce and water, Chaozhou style steamed milkfish is another great way of cooking this kind of fish.

材料

虱目魚肚2片（約600克）、蔥花11/2
大匙

醬料

豆豉11/2大匙、米酒1/2大匙、蒜末1
大匙、蔥末2大匙、嫩薑末2大匙、紅
辣椒末1大匙、高湯1/4杯、醬油2大
匙、細砂糖1/2小匙、香油1/2大匙、
白胡椒粉少許

做法

1. 豆豉切碎；高湯做法參照p.117；醬
   料的材料拌勻。
2. 將虱目魚肚放入盤內，淋上醬料，
   放入蒸籠（水滾了）中，以中大火
   蒸約10分鐘，取出，撒上蔥花。

這樣做更省時

1. 使用休閒鍋或双享鍋烹調，只要將
   虱目魚肚放入鍋內，淋入醬汁，蓋
   上鍋蓋煮，待冒煙後改成小火煮5
   分鐘，移至外鍋燜3～4分鐘即可。
2. 乾豆豉和米酒放入易拉轉中打勻，
   只要手拉一拉很方便。

Ingredients

2 slices of milkfish fillet (about 600g), 11/2T finely chopped green onions

Sauce

11/2T fermented soy beans, 1/2T rice cooking wine, 1T minced garlic, 2T minced green onions, 2T minced tender ginger, 1T minced red chillies, 1/4cup broth, 2T soy sauce, 1/2t granulated sugar, 1/2T sesame oil, white pepper to taste

Directions

1. Chop the fermented soy beans. See p117 for the broth recipe. Mix the sauce well.
2. Place the milkfish fillet in a plate. Pour on the sauce and place in a steamer over boiling water. Steam on medium-high heat for 10 minutes. Remove and sprinkle with finely chopped green onions.

Time-saving method

1. When using a Hotpan or Durotherm, just place milkfish fillets in the pot. Pour in the sauce then put the lid on. Reduce to low heat when the steam begins to hiss out of the pan and cook for 5 minutes. Remove to the outer pot and simmer for another 3 to 4 minutes.
2. You can also mix the dried fermented soy beans and rice cooking wine in Twist & Chop. All you need to do is to pull the string. How convenient!

CC 烹調秘訣 | Cooking tips

1. 如果使用的是乾的豆豉，要先放在米酒中泡約10分鐘再使用。
2. 也可將乾豆豉和米酒放入食物調理機內打勻，至有小顆粒狀，不可太碎，香氣更重。

1. To use dried fermented soy beans, you have to soak in rice cooking wine for 10 minutes in advance.
2. To increase the aroma, put the dried fermented soy beans and rice cooking wine into a food processor and blend until the beans turn into small particles. Do not overblend.

醬淋吳郭魚

Sautéed Tilapia in Soy Sauce

吳郭魚是很容易買到的魚，可清蒸
或煎熟後，搭配糖醋、豆瓣等醬汁
食用，用少少的錢就能享受海鮮。

Tilapia is easy to buy. It can be
steamed or sautéed then be served
with sweet and sour sauce or soy bean
paste. What an inexpensive way to
savor seafood.

材料

吳郭魚1尾（約600克）、水1/2杯、蒜末1大匙、蔥末1大匙、薑末1大匙、太白粉1大匙、鹽少許、白胡椒粉少許、黑醋1/3大匙、香油1/2大匙

調味料

醬油3大匙、細砂糖3大匙

辛香料

紅辣椒末1大匙、蒜苗末1支份量、香菜末3大匙

做法

1. 吳郭魚洗淨，擦乾水分，魚的正反面都撒上少許鹽和白胡椒粉，置於一旁約10分鐘，取出。
2. 鍋燒熱，倒入2大匙油，放入吳郭魚煎至兩面酥脆，取出。
3. 鍋燒熱，倒入1大匙油，先放入蒜末、蔥末和薑末爆香，續入調味料，以小火煮至湯汁起泡，倒入水煮滾，加入黑醋，以太白粉勾芡，再淋上香油拌勻成醬汁。
4. 將辛香料拌勻，均勻地鋪在吳郭魚上面，最後淋上醬汁即可食用。

Ingredients

1 tilapia (about 600g), 1/2 cup water, 1T minced garlic, 1T minced green onions, 1T minced ginger, 1T tapioca starch, salt as needed, white pepper to taste, 1/3 T black vinegar, 1/2 T sesame oil

Seasoning

3T soy sauce, 3T granulated sugar

For aromatics

1T minced red chillies, 1 stalk of minced leek, 3T minced coriander

Directions

1. Rinse the tilapia and pat dry. Sprinkle with a little bit of salt and pepper on both sides. Set aside for 10 minutes. Remove.
2. Heat the pan. Add 2T oil. Sauté the tilapia until both sides are crispy. Remove.
3. Heat the pan. Add 1T oil. Stir-fry the minced garlic, green onions and ginger until fragrant. Add the seasonings and bring to bubbling on low heat. Add water and bring to a boil. Add black vinegar and thicken with tapioca starch. Pour on sesame oil, mix well then the sauce is ready.
4. Combine the spices for aromatics well then evenly sprinkle on the tilapia. Last, pour on the sauce and serve.

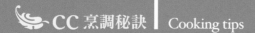

CC 烹調秘訣 | Cooking tips

1. 做法3.中煮醬油和細砂糖時一定要用小火，而且一定要煮至湯汁起泡才會散發出香氣。
2. 除了吳郭魚，鯧魚也很適合烹調這道料理。

1. The soy sauce and granulated sugar in direction no.2 must be cooked on low heat till bubbling to bring out the aroma.
2. In addition to tilapia, pomfret is also an ideal ingredient for this dish.sweetness of sugar.

破布子蒸鱈魚

Steamed Cod with Preserved Cummingcordia

破布子是許多主婦很喜歡的食材，
甘醇的風味，可用來蒸魚、蒸肉，
為原本平淡的食材口味更加分。

Preserved cummingcordia is a very
popular ingredient among housewives.
By steaming with either fish or meat, its
rich flavor can enrich the taste of plain
ingredients.

材料

鱈魚1片、破布子3大匙、蔥花1大匙

醬料

蔥末1大匙、嫩薑末1大匙、紅辣椒末
1/2大匙、米酒1大匙、高湯4大匙、
醬油1大匙、糖少許、香油1/2大匙

做法

1. 高湯做法參照p.117；醬料拌勻。
2. 鱈魚擦乾後放入盤中，將破布子均勻地撒在鱈魚上面，放入蒸籠（水滾了）中蒸約8分鐘，取出。
3. 淋上醬汁，撒上蔥絲即可食用。

這樣做更省時

如果使用休閒鍋或双享鍋，將鱈魚放入鍋內，均勻撒入破布子，淋上醬汁，蓋上鍋蓋加熱，待冒煙改小火煮4分鐘即可。

Ingredients

1 slice of cod, 3T preserved cummingcordia, 1T finely chopped green onions

Sauce

1T minced green onions, 1T minced tender ginger, 1/2T minced red chillies, 1T rice cooking wine, 4T broth, 1T soy sauce, granulated sugar to taste, 1/2T sesame oil

Directions

1. See p.117 for the broth recipe. Mix the sauce well.
2. Pat the cod dry and place on a plate. Sprinkle the preserved cummingcordia evenly on the cod then place in a steamer over boiling water. Steam for 8 minutes. Remove.
3. Pour on the sauce. Sprinkle with shredded green onions and serve.

Time-saving method

When using a Hotpan or Durotherm, place the cod in the pan, evenly sprinkle with preserved cummingcordia and pour on the sauce. Cover and turn on the heat. Reduce to low heat when the steam begins to hiss out. Cook for 4 minutes and serve.

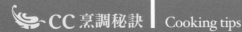

CC 烹調秘訣 | Cooking tips

1. 除了鱈魚，也可以買虱目魚、鱸魚做這道菜。
2. 破布子又叫樹子，甘甜香醇，一般在超市、雜貨店都有賣，多為罐裝。

1. You can substitute milkfish or sea bass over cod.
2. Cummingcordia is also known as Sebastan plum cordia. It has a sweet and mellow flavor. Preserved cummingcordia is usually packaged in jars and can be bought at supermarkets and grocery stores.

沙茶鮮蝦粉絲煲

Shrimp and Green Bean Noodles Casserole
in Shacha Sauce

很多人嗜吃海鮮,加上這道料理的做法簡單、材
料易取得,在我的烹飪課程裡一直很受歡迎。尤
其加入了冬粉,很適合當正餐,宴客也OK。

This dish has been very popular in my class because
many people crave for seafood, plus the dish is easy
to make and ingredients are easy to find. After adding
green bean noodles, it can be served as main dish or an
entertaining meal.

材料

蝦子400克、冬粉2把、蔥末2大匙、蒜末2
大匙、薑末1大匙、紅辣椒末1大匙、沙茶
醬4大匙、米酒2大匙、黑醋1大匙、高湯2
杯、太白粉3大匙、香油1大匙、香菜適量

醃蝦料

米酒1大匙、香油1/2大匙、白胡椒粉少許

調味料

米酒1大匙、醬油11/2大匙、細砂糖1/2小
匙、白胡椒粉少許

做法

1. 蝦子去殼留下尾巴，由背部劃一刀挑除
   腸泥，放入醃蝦料中醃約15分鐘；冬粉
   浸泡水至軟，再剪成適當的長度；高湯
   做法參照p.117。
2. 蝦子沾裹太白粉，放入油溫160℃的油
   鍋中炸，炸至變色後撈出，瀝乾油分。
3. 鍋燒熱，倒入1大匙油，先放入各1大匙
   的蔥末、蒜末、薑末和紅辣椒末爆香，
   續入1大匙的沙茶醬炒一下，放入蝦子，
   淋入米酒稍微炒幾下，取出。
4. 原鍋再燒熱，倒入1大匙油，先放入剩
   餘的蔥末、蒜末爆香，續入3大匙的沙
   茶醬、黑醋和高湯煮滾，放入冬粉燜煮
   一下，再放入做法3.、調味料拌勻燜一
   下，淋上香油即可食用。

Ingredients

400g shrimps, 2 bunches of green bean noodles, 2T minced green onions, 2T minced garlic, 1T minced ginger, 1T minced red chillies, 4T Shacha sauce, 2T rice cooking wine, 1T black vinegar, 2cup broth, 3T tapioca starch, 1T sesame oil, coriander as needed

Marinade

1T rice cooking wine, 11/2T soy sauce, white pepper to taste

Seasonings

1T rice cooking wine, 11/2T soy sauce, 1/2t granulated sugar, white pepper to taste

Directions

1. Peel the shrimps and keep the tails. Slit along the outer curve to expose the sand veins then devein the shrimps. Marinate for 15 minutes. Soak the green bean noodles in water until softened then cut into appropriate sections. See p.117 for the broth recipe.
2. Coat the shrimps with tapioca starch. Deep-fry in oil at 160°C until the color change. Remove and Drain.
3. Heat the pan. Add 1T oil. Stir-fry 1T of minced green onions, minced garlics, minced ginger and minced red chillies first until fragrant. Then add 1T Shacha sauce and stir-fry shortly. Add the shrimps and the cooking wine then stir-fry briefly. Remove.
4. Heat the same pan. Add 1T oil. Stir-fry the remaining minced green onions and garlic until fragrant. Add 3T Shacha sauce, black vinegar and broth then bring to a boil. Add the green bean onions and simmer for a little bit. Add the shrimp mixture in directions no.3 and seasonings. Mix well and simmer for a little bit. Drizzle with sesame oil and serve.

 CC 烹調秘訣 | Tips for a perfect dish

1. 除了蝦子，螃蟹也很適合烹調這道料理。
2. 用XO醬取代沙茶醬，變成一道港式風味的料理。

1. In addition to shrimps, crab is also an ideal ingredient for making this dish.
2. By substituting XO sauce for Shacha sauce, this dish will turn into a Hong King style cuisine.

蜆肉韭黃

Clams and Yellow Chives

這道濃而不膩、清而不淡的潮洲風味料理，是以海鮮為主材料，搭配重口味的豆豉烹調而成。只要掌握好豆豉的鹹度，成功機率非常高。

The flavor of this Hunan-style dish is tasty but not greasy, refreshing but not bland. The ingredients are mainly seafood and are cooked with fermented black bean which is known for its strong flavor. The dish is not that difficult to make as long as you control the saltiness of the fermented black beans.

材料

去骨去皮雞腿肉2隻、蜆600克、米酒4大匙、豆豉1大匙、蒜仁2粒、紅辣椒2支、薑片3片、高湯1/5杯、韭黃120克

醃肉料

鹽1小匙、香油1/2大匙、米酒1大匙、細砂糖1/2小匙、白胡椒粉少許

調味料

鹽、白胡椒粉和細砂糖各少許

做法

1. 雞肉切條狀，放入醃肉料中醃約30分鐘；高湯做法參照p.117。
2. 將蜆放入鍋中，淋入2大匙米酒，蓋上鍋蓋加熱，煮至蜆的殼打開，取出蜆肉，蜆汁則瀝掉雜質備用。
3. 豆豉放入2大匙米酒中浸泡約10分鐘後取出；蒜仁、紅辣椒都切片；韭黃切段。
4. 鍋燒熱，倒入2大匙油，倒入雞肉拌炒至熟後先取出，再放入蒜片、薑片爆香，倒入豆豉炒香，再加入炒熟的雞肉。
5. 淋入高湯和蜆汁，加入調味料，最後倒入韭黃、蜆肉和紅辣椒拌炒幾下，淋上香油即可食用。

Ingredients

2 boneless and skinless chicken quarter legs, 600g clams, 4T rice cooking wine, 1T fermented black beans, 2 cloves of garlic, 2 red chillies, 3 slices of ginger, 1/5 cup of broth, 120g yellow chives

Marinade

1t salt, 1/2T sesame oil, 1T rice cooking wine, 1/2t granulated sugar, white pepper to taste

Seasonings

Salt, white pepper and granulated sugar to taste

Directions

1. Cut chicken into strips. Marinate for 30 minutes. See p.117 for the broth recipe.
2. Place the clams in the pan. Drizzle 2T of rice cooking wine. Put the lid on and turn on the heat until all the shells are open. Remove the meat. Filter out the impurities and reserve the clam juice.
3. Soak the fermented black beans in 2T of rice cooking wine for 10 minutes. Remove. Slice the garlic cloves and red chillies. Cut the yellow chives in sections.
4. Heat the pan. Add 2T of oil. Add the chicken and stir-fry until cooked thoroughly. Transfer the chicken to a plate. Then add the sliced garlic and ginger and stir-fry until fragrant. Add the fermented black beans and stir-fry until fragrant. Last, add the cooked chicken.
5. Drizzle the broth and clam juice. Add the seasonings. Last, add the yellow chives, clam meat and red chillies and stir-fry briefly. Drizzle the sesame oil and serve.

🍃 CC 烹調秘訣 | Cooking tips

1. 韭黃不可以炒得過久，否則會爛爛的導致口感不佳。
2. 注意這道料理不可使用綠韭菜來做，味道會完全不對。
3. 蜆汁和豆豉本身都已經有鹹度，需斟酌加入調味料。

1. Do not overcook yellow chive. Otherwise it will taste mushy.
2. Be aware not to cook this dish with green chive. The flavor is totally different.
3. Clam juice and fermented bean are already tasted salty. Be careful not to overadd the seasonings.

砂鍋魚頭

Fish Head Casserole

即使許多餐廳都有這道菜，但我還
是最喜歡媽媽做的口味，今天就把
媽媽的獨家秘方告訴讀者，和大家
一起品嘗。

This dish is served in many restaurants,
but my mom's version is still my
favorite. Today, I'm going to share her
secret recipe with you so that everyone
can give it a try.

材料

魚頭（鰱魚、鯉魚或鮭魚）1/2個、洋蔥1/2個、大白菜300克、豆腐1塊、魚丸150克、魚板100克、金針菇200克、蛤蜊300克、肉片200克、蒜苗1支、米酒2大匙、蔥4支、薑片6片、紅辣椒2支、地瓜粉適量、香菜適量、高湯2,000c.c.。

調味料

鹽少許、白胡椒粉適量、沙茶醬2大匙、黑醋3大匙、冰糖1大匙、醬油2大匙

做法

1. 洋蔥切絲；豆腐切塊；2支蔥、紅辣椒、蒜苗都切段；高湯做法參照p.117。
2. 取2支蔥和3片薑片拍扁，放入容器中，倒入米酒，用手抓蔥、薑，抓至蔥、薑的汁液滲出，放入魚頭，均勻擦拭蔥、薑汁液，撒上鹽、白胡椒粉，置於一旁30分鐘。
3. 將魚頭均勻沾上地瓜粉，放入油溫160℃的油鍋中炸，炸至呈金黃色後撈出。也可放入烤箱，以上下火230℃烤至熟。
4. 鍋燒熱，倒入1大匙油，先放入洋蔥絲、蔥段、剩餘的薑片和紅辣椒炒香，續入沙茶醬、黑醋、冰糖、鹽、醬油和高湯煮滾，放入大白菜和豆腐煮約5分鐘。
5. 接著加入魚頭煮約5分鐘，再放入魚丸、魚板、金針菇、蛤蜊和肉片，最後撒入蒜苗和香菜即可食用。

Ingredients

1/2 fish head(silver carp, carp or salmon), 1/2 onion, 300g Chinese cabbages, 1 square tofu, 150g fish balls, 100g fish cakes, 200g enoki mushrooms, 300g clams, 200g sliced meat, 1 stalk of leek, 2T rice cooking wine, 4 stalks of green onions, 6 sliced ginger, 2 red chillies, sweet potato flour as needed, coriander as needed, 2,000c.c. broth

Seasonings

Salt to taste, white pepper as needed, 2T Shacha sauce, 3T black vinegar, 1T rock sugar, 2T soy sauce

Directions

1. Shred the onion. Cut the tofu into pieces. Cut 2 stalks of green onions, red chillies and leek into sections. See p.117 for broth recipe.
2. Pressed 2 stalks of green onions and 3 sliced ginger. Put in a mixing bowl. Pour in the rice cooking wine. Toss the green onions and ginger by hand until the juice comes out. Add the fish head and rub it with the juice of ginger and green onions thoroughly. Sprinkle with salt and white pepper. Set aside for 30 minutes.
3. Coat the fish head evenly with sweet potato flour. Deep-fry in oil at 160°C until golden brown and remove. The alternative method is to bake in the oven at 230°C until cooked thoroughly.
4. Heat the pan. Add 1T oil. Stir-fry the shredded onions, green onions sections, the remaining sliced ginger and red chillies first until fragrant. Then add the Shacha sauce, black vinegar, rock sugar, salt, soy sauce and broth and bring to a boil. Add the Chinese cabbages and tofu and cook for 5 minutes.
5. Add the fish head and cook for 5 minutes. Then add the fish balls, fish cakes, enoki mushrooms, clams and sliced meat. Last, sprinkle with leek and coriander. Serve.

CC 烹調秘訣 | Cooking tips

1. 先以蔥、薑片和米酒抓勻再擦拭魚頭，可去掉腥味、增添香氣。
2. 油炸魚類料理後，餘油會有腥味，但倒掉又很可惜，建議將蔥尾、薑皮放入炸過魚的油裡面炸，炸至蔥變成焦黃色後取出，這樣油可以去掉腥味，就能用來烹調其他料理了。

1. Mixing green onions and sliced ginger with rice cooking wine and rubbing on the fish head can remove the fish odor and increase the aroma.
2. After deep-frying seafood, the leftover oil will be smelly, but it's such a waste to just discard it. There is a way to solve the problem. You can dump the green onions roots and ginger skins into the leftover oil and deep-fry until the green onions turn dark brown. Thus, the odor of oil will be eliminated and can be reused to cook other dishes.

Part 3 肉類

餐桌上的鎂光燈焦點，就是色香味俱全的肉類了！
想想，要是少了一道主菜，
縱然有再精緻的各式菜餚，總也顯得黯然失色；
相反的，只要擺上一道精心烹調的肉類料理，
就算桌上只是三菜一湯的簡單小菜，
頓時就會增色不少。

With their great appearance, aroma and flavor,
meat dishes are always in the spotlight on dining tables.
Just think about it. No matter how exquisite a meal is,
it would somehow look pale without the main dish.
On the contrary, once an elaborately made meat dish is served,
even an ordinary meal with three dishes and a soup would look just wonderful.

檸檬香味雞

Lemon Chicken

以清新的檸檬香入菜，微微酸甜的
滋味，是一道非常清爽、百吃不膩
的料理。

With the refreshing lemon fragrance
and slightly sweet-and-sour flavor, this is
a very light dish that you will never get
tired of it.

材料

去骨雞腿3隻、地瓜粉1/4杯、太白粉1/4杯、檸檬皮絲適量

醃肉料

米酒1大匙、鹽1/2大匙、香油1大匙、細砂糖1/2小匙、薑汁1大匙、蒜粉1小匙

檸檬醬汁

高湯1/4杯、米酒1大匙、細砂糖1/3大匙、檸檬汁31/2大匙、鹽適量、太白粉（勾芡用）11/2大匙

做法

1. 雞腿肉切塊狀，放入醃肉料中醃約1小時；地瓜粉、太白粉混合；高湯做法參照p.117。
2. 雞肉塊均勻沾上粉類，放入油溫160℃的油鍋中炸，炸至呈金黃色，撈出瀝乾油分，盛盤。
3. 將檸檬醬汁的材料倒入小鍋中煮滾，以太白粉勾芡，淋在炸好的雞肉塊上面，撒上檸檬皮絲裝飾即可食用。

Ingredients

3 boneless Chicken thighs, 1/4cup sweet potato flour, 1/4cup tapioca starch, lemon zest as needed

Marinade

1T rice cooking wine, 1/2T salt, 1T sesame oil, 1/2t granulated sugar, 1T ginger juice, 1t garlic powder

Lemon Sauce

1/4cup broth, 1T rice cooking wine, 1/3 T granulated sugar, 31/2T lemon juice, salt as needed, 11/2T tapioca starch (for thickening)

Directions

1. Cut the chicken thighs into bite-size pieces. Marinate for 1 hour. Mix the sweet potato flour and tapioca starch. See p.117 for the broth recipe.
2. Coat the chicken with flour mixture evenly. Deep-fry in the oil at 160°C until golden brown. Remove and drain. Place on a serving plate.
3. Pour the lemon sauce in a small saucepot. Bring to a boil and thicken with tapioca starch. Drizzle on the deep-fried chicken. Garnish with lemon zest and serve.

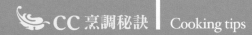

CC 烹調秘訣 | Cooking tips

1. 雞塊放入油鍋中炸時，不要馬上翻動，等約1分鐘30秒，讓食物在油鍋內穩定，皮才不易脫落。而且油炸時，火不能大火，以中小火炸，否則火太大，會使食物外皮已經焦黃但肉內還是生的，待要起鍋時才改成中大火。
2. 在炸好的雞塊上面撒胡椒粉、辣椒粉和稍微炸一下的九層塔拌勻，就變成另一種風味的鹹酥雞了。

1. When deep-frying chicken, do not flip the meat immediately. Wait for one and half minutes to let the meat become stable in the pan so that the skin will not fall off easily. Never deep-fry food on high heat. Only on medium-low heat. If the heat is too high, it will burn the outside part of meat while the inside part remains uncooked. Only turn on to medium-high heat when it's almost done.
2. If you sprinkle with peppers, red chillies and mix the cooked chicken with slightly deep-fried basil, you will have Taiwanese fried salty chicken.

香蔥翡翠雞

Emerald Chicken with Green Onions

這是一道兼具色香味的美食，當成家常、宴客菜都很合適，是廣受我烹飪教室學生喜愛的好菜。

This dish is perfect in color, aroma and taste. This is one of the most popular dishes in my cooking class and is good to be cooked as either everyday meal or an entertainment dish.

材料

去骨雞腿3隻、美生菜或高麗菜100克、蔥2
支、紅辣椒1支

醃肉料

蔥2支、薑片3片、醬油3大匙、香油1大匙、
米酒1大匙、細砂糖1/3小匙、五香粉1/3小
匙、白胡椒粉少許、香菜頭3支

醬汁

香菜末3大匙、蔥末2大匙、嫩薑末1大匙、醬
油膏4大匙、細砂糖1大匙、香油1大匙、檸檬
汁11/2大匙

做法

1. 生菜剝成片後放入冰水中浸泡，冰鎮約15
   分鐘後切絲；材料中的蔥和紅辣椒切絲；
   醃肉料中的蔥、薑片拍扁；醬汁拌勻。
2. 將醃肉料放入調理盆中，戴上手套後用力
   抓勻，可使蔥、薑、香菜頭的香味更完整
   滲出，然後將雞腿放入拌勻，醃約1小時。
3. 鍋燒熱，倒入1大匙油，放入雞腿，皮朝鍋
   底煎約1分鐘30秒，然後翻面煎至全熟，取
   出切塊狀。也可放入上下火230℃的烤箱中
   15～20分鐘至熟，但仍需視自家烤箱斟酌
   時間。
4. 將生菜鋪在盤中，放上雞腿，淋上醬汁，
   最後撒上蔥絲和紅辣椒絲即可食用。

這樣做更省時

如果使用休閒鍋或双享鍋煎雞腿，只要鍋熱倒
入1大匙油，將雞腿皮朝鍋底煎約2分鐘，翻
面，蓋上鍋蓋以小火煮6～7分鐘即可。

Ingredients

3 boneless chicken leg quarters, 100g lettuces or
Taiwanese cabbages, 2 stalks of green onions, 1 red
chillies

Marinade

2 stalks of green onions, 3 slices of ginger, 3T soy
sauce, 1T sesame oil, 1T rice cooking wine, 1/3 t
granulated sugar, 1/3 t Five Spices Powder, white
pepper as need, 3 coriander roots

Sauce

3T minced coriander, 2T minced green onions,
1T minced tender ginger, 4T soy sauce paste, 1T
granulated sugar, 1T sesame oil, 11/2T lemon juice

Directions

1. Tear the lettuces into pieces and soak in
   ice water for 15 minutes. Drain and shred.
   Shred the green onions and red chillies in the
   ingredients. Press the green onions and the
   ginger slices in the marinade. Mix the sauce
   well.
2. Pour the marinade into a bowl. Put on the
   gloves and mix it well to bring out the flavor of
   ginger and coriander roots. Add the chicken
   quarter legs and mix well. Marinate for 1 hour.
3. Heat the pan. Add 1T of oil. Add the chicken
   with skin side down and sear for one and half
   minutes. Flip to the other side and pan-fry until
   cook thoroughly. Remove and chop into pieces.
   You can also bake it in the oven at 230°C until
   cook thoroughly. It takes about 15 to 20 minutes
   but the exact cooking time varies, depending on
   oven.
4. Put the lettuce on a serving plate. Arrange the
   chicken quarter legs. Drizzle the sauce and
   sprinkle the shredded green onions and red
   chillies . Serve.

Time-saving method

When pan-frying the chicken quarter legs in a
Hotpan or Durotherm, you just need to add 1T of
oil after the pan is hot. Then sear the chicken with
skin side down for 2 minutes. Flip. Put the lid on
and reduce to low heat for 6 to 7 minutes.

CC 烹調秘訣 | Cooking tips

1. 這道菜搭配的醬汁也能用來拌麵，非常可口。
2. 新鮮生菜泡過冰水後再切絲，可使生菜的口感更脆。

1. You can mix the sauce of this dish with noodles. It's very yummy.
2. Soaking fresh lettuces in ice water before tearing into pieces can make it taste crunchier.

湘味蒜辣雞丁

Hunan-Style Spicy Chicken with Garlic

這道常出現在我家餐桌上的湖南味料理，是喜歡吃嗆辣口味的人不能錯過的。一聞到香味，肚子馬上開始咕咕叫，恨不得立刻開飯。

I cook this Hunan-style dish very often at home. Don't miss this one if you are a spicy food lover. The aroma of this yummy dish always makes my stomach start rumbling and can't wait to eat it.

材料

去皮去骨雞腿3隻、紅辣椒6支、蒜苗1支、蒜仁12粒、豆豉3大匙、米酒3大匙

醃肉料

醬油2大匙、香油1大匙、細砂糖1/2小匙、米酒1大匙、白胡椒粉少許

調味料

白胡椒粉少許、香油1大匙

做法

1. 雞腿肉切約3×3公分的塊狀，放入醃肉料中醃約30分鐘；紅辣椒和蒜苗都切圈狀；蒜仁切片；豆豉放入2大匙米酒中浸泡約10分鐘後取出。
2. 將雞腿肉放入油溫160℃的油鍋中炸，炸至呈金黃色後撈出。
3. 鍋燒熱，倒入2大匙油，先放入蒜片爆香，續入豆豉炒香，淋入1大匙米酒，加入雞腿肉和紅辣椒拌炒一下，撒入白胡椒粉，加入蒜苗快炒幾下，淋入香油即可食用。

這樣做更省時

油炸雞腿肉時，也可用材質好的不沾鍋燒。只要將不沾鍋燒熱，倒入1 1/2大匙油，放入雞腿肉煎至成焦黃色，更方便。

Ingredients

3 boneless and skinless chicken quarter legs, 6 red chillies, 1 leek, 12 garlic cloves, 3T fermented black beans, 3T rice cooking wine

Marinade

2T soy sauce, 1T sesame oil, 1/2t granulated sugar, 1T rice cooking wine, white pepper to taste

Seasonings

White pepper to taste, 1T sesame oil

Directions

1. Cut the chicken into 3x3cm thick pieces. Marinate for 30 minutes. Slice the red chillies and leek. Slice the garlic clove. Soak the fermented black beans in 2T of rice cooking wine for 10 minutes. Remove.
2. Deep-fry the chicken in oil at 160°C until golden brown. Remove.
3. Heat the pan. Add 2T of oil. First, stir-fry garlic slices until fragrant. Then add the fermented black beans and stir-fry until fragrant. Add 1T of rice cooking wine, chicken and red chillies and stir–fry quickly. Sprinkle with white pepper. Add leek and stir-fry briefly. Drizzle the sesame oil and serve.

Time-saving method

You can deep-fry chicken quarter legs on a good non-stick skillet. It's more convenient. Just heat the pan, add 1 1/2T of oil and fry the chicken legs until brown.

CC 烹調秘訣 | Cooking tips

將雞肉改成切片的五花肉放入鍋中煎至焦黃，不需先醃過，但其他材料相同，就變成湖南口味肉片了。愛重口味的人，建議加入乾辣椒。

You can substitute sliced pork belly with chicken. You don't need to marinate the meat in advance. Just sear the pork belly until brown and cook with the same remaining ingredients. Then you will have a dish of Hunan-style sliced pork. If you crave for strong flavors, you would probably like to add dried red chillies.

宮保雞丁

Kung Pao Chicken

這是道四川的名菜，宮保就是乾辣椒，具有特殊香氣和辣味，拿來做菜最能促進食慾，多吃好幾碗飯。

This is a famous dish in Sichuan cuisine. Kung Pao refers to dried chilies which provide a hint of tang and the heat. Cooking with this spice will stimulate people's appetite and make them eat several bowls of rice.

材料

去皮去骨雞胸肉500克、乾辣椒30克、脆熟花生80克、蔥2支、薑片5片、蒜仁5粒、白醋1/3大匙

醃肉料

醬油21/2大匙、細砂糖1/3小匙、米酒1大匙、香油1/2大匙、白胡椒粉少許、雞蛋1個、太白粉3大匙

調味料

醬油2大匙、細砂糖1/2小匙、米酒1大匙、香油1/2大匙、白醋1/3大匙、高湯3大匙

做法

1. 蔥切段；雞蛋打散；蒜仁切片；高湯做法參照p.117；調味料拌勻。
2. 雞胸肉切塊狀，放入醃肉料中拌勻，加入蛋液拌勻，再放入太白粉拌勻，醃約30分鐘。
3. 將雞胸肉放入油溫160℃的油鍋中炸，炸至呈金黃色後撈出，不要炸太老。
4. 鍋燒熱，倒入1大匙油，先放入蔥、薑和蒜爆香，續入乾辣椒，炒至乾辣椒的香氣充分散出，加入雞胸肉和調味料，快速拌炒均勻，淋入鍋邊醋（白醋），最後淋入香油。
5. 將脆熟花生鋪在盤底，放上炒好的做法4.即可食用。

Ingredients

500g skinless and boneless chicken breast, 30g dried chilies, 80g pre-roasted peanuts, 2 stalks of green onions, 5 slices of ginger, 5 garlic cloves, 1/3 T white vinegar

Marinade

21/2T soy sauce, 1/3 t granulated sugar, 1T rice cooking wine, 1/2T sesame oil, white pepper to taste, 1 egg, 3T tapioca starch

Seasonings

2T soy sauce, 1/2t granulated sugar, 1T rice cooking wine, 1/2T sesame oil, 1/3 T white vinegar, 3T broth

Directions

1. Cut the green onions in section. Whisk the egg. Slice the garlic cloves. See p. 117 for broth recipe. Mix the seasonings well.
2. Cut the chicken breast into pieces,and marinate,mix wisked egg and tapiocal starch well.Marinate for 30 minutes.
3. Deep-fry the chicken breast in oil at 160°C until golden brown. Be aware not to overcook.
4. Heat the pan. Add 1T of oil. Stir-fry the green onions, ginger and garlics first until fragrant. Then add the dried chilies until fully fragrant. Add the chicken breast and seasonings. Stir-fry quickly. Drizzle the white vinegar along the rim of pan. Last, drizzle the sesame oil.
5. Put pre-roasted peanuts on the plate then top with the chicken mixture in direction **4.** Serve.

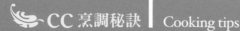
CC 烹調秘訣 Cooking tips

除了油炸雞胸肉以外，也可以用煎的。將鍋燒熱，倒入3大匙油，放入雞胸肉拌炒至快熟的狀態就先取出。

In addition to deep-frying, the chicken breast can also be sautéed. Heat the pan. Add 3 T oil of oil. Stir-fry the chicken breast until it's almost done and remove for later use.

荷香蒸雞

Steamed Chicken with Lotus Leaf

這道料理很適合在大節日時烹調。
荷葉天然的香氣完全滲入雞肉裡，
帶有特殊香氣，當然也可以做荷葉
鴨、荷葉肉喔！還可以將整隻荷香
蒸雞用來祭祀喔！

This dish is ideal to serve in big holidays.
The chicken absorbs all the distinctive
natural aroma of lotus leaf. You can also
steam ducks and pork with lotus leaves!
Even steam the whole chicken with
lotus leaves for worshiping.

材料

土雞或仿土雞1隻、荷葉1張、沙拉筍100克、乾香菇8朵、火腿100克、肉絲100克、蔥2支、嫩薑50克、年糕紙1張（玻璃紙）

醃雞肉料

干貝蠔油3大匙、醬油2大匙、醬油膏2大匙、紹興酒2大匙、白胡椒粉1/2小匙、香油1大匙、拍扁的蔥2支、拍扁的薑片3片

醃肉料

醬油1/2大匙、米酒1/3大匙、香油1小匙、白胡椒粉少許

做法

1. 土雞洗淨，瀝乾水分，放入醃雞肉料中醃約2小時；乾香菇泡水至軟後切絲；沙拉筍切絲；火腿切絲；肉絲放入醃料中醃約15分鐘；蔥和嫩薑都切絲。
2. 荷葉洗淨，放入滾水中汆燙一下，取出再以清水洗淨，瀝乾水分。
3. 先取適量的香菇、沙拉筍、火腿和肉絲塞入雞的肚子內，再塞入少許蔥和嫩薑。
4. 將年糕紙鋪平，先鋪上荷葉，放上整隻雞，在雞全身均勻地鋪滿香菇、沙拉筍、火腿和肉絲，整個以荷葉包好，放入蒸籠（水滾了）蒸2小時30分鐘即可食用。

這樣做更省時

除了蒸籠外，如果使用壓力鍋燉煮，鍋中倒入800c.c.的水，放入荷葉包好的整隻雞加熱，等上升兩條紅線，再改小火煮約50分鐘即可。

Ingredients

1 native chicken or simulated native chicken, 1 lotus leaf, 100g bamboo shoot, 8 dried shitake mushrooms, 100g ham, 100g shredded pork, 2 stalks of green onions, 50g tender ginger, 1 sheet of cellophane

Chicken marinade

3T oyster sauce with dried scallops, 2T soy sauce, 2T soy sauce paste, 2T Shaohsing wine, 1/2t white pepper, 1T sesame oil, 2 stalks of pressed green onions, 3 sliced of pressed ginger

Meat marinade

1/2T soy sauce, 1/3 T rice cooking wine, 1t sesame oil, white pepper to taste

Directions

1. Wash the native chicken. Drain. Marinade for 2 hours. Soak the dried shitake mushrooms until softened then shred. Shred the bamboo shoots. Shred the ham. Soak the shredded pork in the meat marinade for 15 minutes. Shred green onions and tender ginger.
2. Wash the lotus leaf. Blanch quickly in boiling water. Remove and wash again. Drain.
3. Stuff enough amount of shitake mushrooms, bamboo shoots, ham and shredded pork into the chicken. Then stuff a little bit of green onions and ginger.
4. Place the cellophane on a flat surface. First, spread out the lotus leaf. Then put the whole chicken on it. Evenly sprinkle shitake mushrooms, bamboo shoots, ham and shredded pork on the chicken. Wrap it with lotus leave. Put it in a steamer on boiling water. Steam for 2 hours and 30 minutes. Serve.

Time-saving method

When using a Duromatic, just add 800c.c. of water and put in the whole chicken wrapped in the lotus leaf. When two red bars are visible, reduce to low heat for 50 minutes. Serve.

 CC 烹調秘訣 | Cooking tips

1. 如果不使用全雞，也可以改成雞腿或半雞製作。
2. 荷葉中間的粗梗要先修剪掉，才不會突出來，但修剪後中間會有小洞，可以將荷葉折起約5公分蓋住小洞即可。
1. You can substitute chicken quarter legs or half chicken for whole chicken.
2. Trim the rough stem of the leaf so that it will not stick out. Although there will be a tiny hole in the center after trimming, you can fold up the leaf about 5cm to cover the hole.

怪味雞

Cold Chicken Appetizer with Hot Sauce

這是四川的名菜之一，醬汁是讓這
道菜好吃的關鍵！醬汁還可以拿來
拌麵喔，一種醬多種吃法。

This is a famous Sichuan cuisine. The
secret of this dish is the sauce! The
sauce can be used in other dishes, such
as mixing with noodles.

材料

去皮去骨雞胸肉400克、蘆筍300克、蔥2支、薑片2片、米酒1大匙

醬汁

白芝麻醬2大匙、醬油1大匙、醬油膏1大匙、細砂糖1/3大匙、辣油1/2大匙、香油1大匙、黑醋1/2大匙、白醋2/3大匙、花椒粉1/2小匙、薑末1大匙、蔥末2大匙、蒸雞肉的湯汁3大匙

調味料

鹽少許、白胡椒粉少許

做法

1. 蔥切段；醬汁拌勻。
2. 雞胸肉放在盤中，淋入米酒，撒上調味料，鋪上蔥段和薑片，放入蒸籠（水滾了）蒸10～12分鐘，取出，待雞胸肉涼後切片。
3. 蘆筍放入滾水中汆燙至熟，取出瀝乾水分後切段，鋪在盤中。
4. 將雞胸肉鋪在蘆筍上面，淋上醬汁即可食用。

這樣做更省時

使用休閒鍋或双享鍋烹調雞胸肉，只要加入4大匙的水，蓋上鍋蓋加熱，待冒煙後改成小火煮7～8分鐘即可。

Ingredients

400g boneless and skinless chicken breast, 300g asparagus, 2 stalks of green onions, 2 slices of ginger, 1T rice cooking wine

Sauce

2T white sesame sauce, 1T soy sauce, 1T soy sauce paste, 1/3T granulated sugar, 1/2T chili oil, 1T sesame oil, 1/2T black vinegar, 2/3T white vinegar, 1/2t Sichuan peppercorn powder, 1T minced ginger, 2T minced green onions, 3T juice left from steaming chicken

Seasonings

Salt to taste, white pepper to taste

Directions

1. Cut the green onions in sections; mix the sauce well.
2. Place the chicken breast in a pan. Pour the rice cooking wine over. Sprinkle with seasonings. Add the green onions sections and sliced ginger. Place in the steamer with boiling water for 10 to 12 minutes. Remove. Let cool and slice.
3. Blanch the asparagus in boiling water until cooked thoroughly. Drain and cut into sections. Arrange on a serving plate.
4. Arrange the chicken breast on the asparagus. Pour on the sauce and serve.

Time-saving method

When cooking chicken breast in a Hotpan or Durotherm, just add 4T of water. Put the lid on and turn on the heat. When the steam hisses out of the pan, reduce to low heat and cook for 7 to 8 minutes.

CC 烹調秘訣 | Cooking tips

可利用祭祀過的水煮雞或雞腿肉取代雞胸肉；小黃瓜取代蘆筍。

You can substitute the boiled chicken or chicken quarter legs used for worshiping over chicken breast. Asparagus can be substituted with cucumbers.

銀芽雞柳

Stir-Fried Chicken Strips with Silver Sprouts

不是每餐都得有重口味的大魚大肉料理，偶爾來道爽口小菜平衡一下吧！清淡的雞胸肉快炒可口的銀芽，讓雞胸肉變得更美味。

It's not necessary to have meat and fish every meal. Sometimes a refreshing dish is a good choice of eating hearty. By stir-frying the crunchy silver sprouts with the light chicken breast, it can bring out the delicate flavor of the meat.

材料

去皮去骨雞胸肉300克、銀芽150克、紅辣椒2根、青椒1/2個、蔥1支、蒜末1大匙、地瓜粉1/3杯、香油1/2大匙

醃肉料

鹽1小匙、香蒜粉1/2小匙、米酒1/2大匙、香油1大匙、攪勻蛋白1個

調味料

高湯3大匙、白胡椒粉少許、香油1小匙、鹽1/2小匙、細砂糖1/3小匙

做法

1. 雞胸肉切條狀，放入醃肉料中醃約30分鐘，取出均勻沾上地瓜粉。
2. 紅辣椒、青椒都切絲；蔥切段；高湯做法參照p.117。
3. 鍋中倒入適量的油，待油熱後放入雞胸肉煎炸至呈金黃色，取出瀝乾油分。
4. 另一鍋燒熱，倒入1大匙油，先放入蔥段、蒜末爆香，續入銀芽、紅辣椒和青椒拌炒，倒入調味料拌勻，再加入雞胸肉快炒幾下，淋入香油即可食用。

Ingredients

300g boneless & skinless chicken breast, 150g bean sprouts, 2 red chillies, 1/2 green pepper, 1 green onion, 1T minced garlic, 1/3 cup of sweet potato flour, 1/2 T sesame oil

Marinade

1t salt, 1/2 t garlic powder, 1/2 T rice cooking wine, 1T sesame oil, 1 whisked egg white

Seasonings

3T broth, white pepper as need, 1t sesame oil, 1/2 t salt, 1/3 t granulated sugar

Directions

1. Cut the chicken breast into strips. Marinate for 30 minutes. Remove and coat evenly with the sweet potato flour.
2. Shred the red chillies and green peppers. Cut green onion into sections. See p.117 for the broth recipe.
3. Add the oil as needed in the pan. Heat the oil and stir-fry the chicken strips until golden brown. Remove and drain.
4. Heat another pan. Add 1T of oil. First stir-fry green onion sections and minced garlic until fragrant. Then stir in the bean sprouts, chili and green peppers. Add the seasonings. Mix well. Last, add the chicken breast and stir quickly and briefly. Drizzle with the sesame oil and serve.

CC 烹調秘訣 | Cooking tips

1. 銀芽是指豆芽菜捏去頭尾。
2. 事先拌勻調味料再倒入鍋中烹調，可節省烹調的時間，避免手忙腳亂。

1. Silver sprouts refer to sprout beans after the ends are being snapped off.
2. By combining the seasonings in advance before adding into the pan, it can not only save your time but also let you cook at ease.

沙茶燜鴨

Braised Duck in Shacha Sauce

沙茶醬源自於福建,當地很多人用來烹調料理,其中尤以這道沙茶燜鴨更是有名!

Shacha sauce was originated in Fuijan. Many local people use it in cooking. This braised duck in Shacha sauce is especially famous.

材料

光鴨1/2隻、小朵乾香菇12朵、蔥2
支、薑片3片、紹興酒1/4杯、高湯2
杯、蒜末1/2大匙、紅辣椒末1大匙、
沙茶醬2 1/2大匙

調味料

醬油1 1/2大匙、醬油膏2大匙

做法

1. 光鴨剁塊狀；乾香菇泡水至軟取
   出；蔥切段；高湯做法參照p.117。
2. 將蔥、薑和鴨肉放入滾水中煮約5
   分鐘，取出鴨肉。
3. 鍋燒熱，倒入2大匙油，先放入蒜
   末、紅辣椒末爆香，續入鴨肉，淋
   入紹興酒，待酒精揮發後加入沙茶
   醬炒香，倒入調味料、高湯煮約40
   分鐘，再放入香菇煮約3分鐘即可
   食用。

Ingredients

1/2 whole duck, 12 small dried shitake
mushrooms, 2 stalks of green onions, 3 slices
of ginger, 1/4 cup of Shaohsing wine, 2 cups of
broth, 1/2T minced garlic, 1T minced red chillies,
2 1/2T Shacha sauce

Seasonings

1 1/2T soy sauce, 2T soy sauce paste

Directions

1. Chop the whole duck into pieces; soak
   the dried shitake mushrooms in water until
   softened. Cut the green onions in sections. See
   p.117 for the broth recipe.
2. Cook the green onions, ginger and the duck in
   boiling water for 5 minutes. Remove the duck.
3. Heat the pan. Add 2T of oil. First, stir-fry minced
   garlic and minced red chillies until fragrant.
   Then add the duck meat and Shaohsing wine.
   Cook until the alcohol evaporates thoroughly.
   Next, add Shacha sauce and stir-fry until
   fragrant. Add seasonings and broth. Cook for
   40 minutes. Last, add shitake mushrooms and
   cook for 3 minutes. Serve.

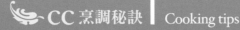

CC 烹調秘訣 | Cooking tips

可以改用雞肉來烹調。以紹興酒取代米酒，更能提升香氣。

You can substitute chicken for duck. Cooking this dish with Shaohsing wine instead of rice
cooking wine can enhance the aroma.

梅子蒸小排
Steamed Spareribs with Plums

這是一道帶有梅香，酸甜口味的下飯菜。尤其在天氣炎熱時，更能促進食慾，白飯一碗又一碗。

This sweet-and-sour dish with plum flavor goes well with rice. It can definitely stimulate your appetite in the hot weather and make you unknowingly eat several bowls of rice.

材料

小排600克、蔥花1大匙、太白粉2大匙、紅辣椒末1大匙

醃肉料

梅子10顆（去籽剁碎）、梅子汁1大匙、醬油11/2大匙、醬油膏11/2大匙、米酒1大匙、白胡椒粉少許、香油1大匙、蕃茄醬11/2大匙、蒜末1大匙、蔥末1大匙

做法

1. 小排剁小塊；梅子去籽剁碎。
2. 小排放入醃肉料中醃約1小時，放入太白粉拌勻。
3. 將醃好的小排放入蒸籠（水滾了）蒸50～60分鐘，撒上蔥花和紅辣椒末即可食用。

這樣做更省時

如果使用壓力鍋蒸煮，可在鍋內放1,000c.c.的水，將小排放入淺蒸盤扣入壓力鍋內，蓋上鍋蓋，加熱後待上升兩條紅線，再改小火蒸約18分鐘，撒上蔥花即可。

Ingredients

600g spareribs, 1T finely chopped green onions, 2T tapioca starch,1T minced red chillies

Marinade

10 plums(chopped, seedless), 1T plum juice, 11/2T soy sauce, 11/2T soy sauce paste, 1T rice cooking wine, white pepper as needed, 1T sesame oil, 11/2T ketchup, 1T minced garlic, 1T finely chopped green onions

Directions

1. Cut the spareribs into small pieces. Pit and chop the plums.
2. Marinate the spareribs for an hour and mix them with tapioca starch.
3. Place the marinated spareribs into a steamer over boiling water for 50 to 60 minutes. Sprinkle with finely chopped green onions and red chillies Serve.

Time-saving method

If you happen to have a Duromatic, add 1,000 c.c. of water to the cooker. Place the ribs on a shallow steam tray and put it into the cooker. Lock the lid. When the steam begins to hiss out of the cooker and two red bars are visible, reduce the heat to low. Cook for about 18 minutes. Sprinkle with the shredded green onions. Serve.

CC 烹調秘訣 | Cooking tips

我習慣用自己醃的梅子來做，一般店家也有賣玻璃瓶裝的，不過並非梅子蜜餞。尤其每年四月梅子盛產，許多人都會大量醃製，梅子汁千萬別浪費了。

I like to use the homemade preserved plum juice. You can find glass-bottled juice in grocery stores, but they are not made from preserved plums. Plums are at the peak of season in every April, many people will preserve lots of them at that time. Don't waste the plum juice.

回鍋肉

Double-Cooked Pork Slices

許多人認為拜拜後的三牲很難處理，每次都是肉片炒醬油，總變不出新花樣。只要改變一下烹調方式，拜拜後的五花肉也能成為經典的四川名菜。

Many people always get headache when brainstorming for the recipes with the sacrifices after worshiping. It seems that the only way of it is to stir-fry the meat slices in soy sauce. Actually, if you could just change the way of cooking it, you can turn the pork belly used in worshiping into a famous Sichuan cuisine.

材料

煮熟的五花肉300克、高麗菜80克、青椒1個、豆乾6片、蒜仁3粒、蔥1支、紅辣椒2支

調味料

辣豆瓣醬2大匙、甜麵醬1 1/2大匙、米酒1 1/2大匙、醬油2 1/2大匙、細砂糖1小匙、香油1小匙

做法

1. 五花肉、豆乾和蒜仁都切片狀；高麗菜用手撕成片狀；青椒和紅辣椒切塊狀；蔥切段。
2. 鍋燒熱，倒入1大匙油，先放入五花肉爆香至呈金黃色，取出。續入豆乾，炒至散發出香氣，取出。
3. 原鍋放入蒜片、蔥段爆香，倒入辣豆瓣醬、甜麵醬，改小火炒至散發出香氣，淋入米酒拌勻，加入五花肉、豆乾以中火拌炒，再加入高麗菜、青椒和紅辣椒快炒。
4. 倒入醬油、細砂糖炒勻，淋入香油即可食用。

Ingredients

300g cooked pork belly, 80g Taiwanese cabbages, 1 green pepper, 6 pieces of dried tofu, 3 cloves of garlic, 1 stalk of green onion, 2 red chillies

Seasonings

2T spicy soybean paste, 1 1/2T sweet soybean paste, 1 1/2T rice cooking wine, 2 1/2T soy sauce, 1t granulated sugar, 1t sesame oil

Directions

1. Slice the pork belly , garlic and dried tofu. Tear the Taiwanese cabbages by hands into pieces. Cut the green pepper and red chillies into bite-size pieces. Cut green onion into sections.
2. Heat the pan. Add 1T of oil. Frist, stir-fry the pork belly until it turns golden. Remove. Then add tofu and stir-fry until fragrant. Remove.
3. Use the same pan. Stir-fry sliced garlic and green sections. Add the spicy soybean paste and sweet soybean paste and stir-fry over low heat until fragrant. Add rice cooking wine and mix well. Stir-in the pork belly and dried tofu on medium heat. Add the cabbages, green and red chillies and stir-fry quickly.
4. Stir in the soy sauce and granulated sugar. Drizzle the sesame oil and serve.

CC 烹調秘訣 | Cooking tips

除了較肥膩的五花肉，也可改成胛心肉、小里肌肉烹調。

Because pork belly is fattier, you can replace it with pork shoulder or tenderloin.

紅燒獅子頭

Braised Lion's Head Meatballs

經典名菜之一，依材料，以及清燉、清蒸和紅燒等烹調方式的差異，呈現出不同風味。我的獅子頭不需經過油炸、煎的過程，清淡不油膩，家常或宴客最優選擇。

One of classic cuisine! The flavor is various, depending on ingredients and cooking styles, such as stew, steam and braise. The lion's head meatballs that I make do not need to be deep-fried or sautéed. The taste is light and not greasy at all. This dish is a perfect choice for both everyday cooking and entertaining.

材料

絞肉600克、蝦米4大匙、荸薺100克、饅頭1個、蔥末3大匙、嫩薑末1大匙、米酒5大匙、高湯1,000c.c.、太白粉3大匙、雞蛋1個、大白菜600克、蔥1支

調味料

醬油11/2大匙、醬油膏11/2大匙、白砂糖1/3小匙、米酒1大匙、香油1大匙、白胡椒粉適量、鹽1小匙

煮汁

醬油3大匙、素蠔油2大匙、八角1粒、冰糖1大匙、甘草1片

做法

1. 蝦米放入3大匙的米酒中浸泡約10分鐘後取出；荸薺壓扁除去汁液後剁碎；饅頭剝碎，放入1/4杯高湯中浸泡約5分（不要太濕，只需讓饅頭吃進高湯）；蔥切斜段。
2. 取一半的蝦米剁碎，放入容器中，倒入絞肉、蔥末和嫩薑末，續入蛋液、調味料、浸泡過的饅頭、荸薺和太白粉，拌勻成有黏性的肉餡。
3. 鍋燒熱，倒入1大匙油，先放入剩餘的蝦米爆香，續入蔥段爆香，立刻放入大白菜拌炒幾下，淋入2大匙的米酒煮一下，倒入煮汁和高湯，改小火煮。
4. 將肉餡做成一顆顆圓球狀，全部放入做法3.內，蓋上鍋蓋，以小火燉燒約1小時20分鐘即可食用。

這樣做更省時

如果使用休閒鍋或双享鍋烹煮肉丸，蓋上鍋蓋，只要加熱約50分鐘即可。

Ingredients

600g ground pork, 4T dried shrimps, 100g water chestnuts, 1 steam bun, 3T minced green onions, 1T minced tender ginger, 5T rice cooking wine, 1,000c.c. broth, 3T tapioca starch, 1 egg, 600g Chinese cabbage, 1 stalk of green onions

Seasonings

11/2T soy sauce, 11/2T soy sauce paste, 1/3 t white granulated sugar, 1T rice cooking wine, 1T sesame oil, white pepper as needed, 1t salt

Cooking sauce

3T soy sauce, 2T vegetarian oyster sauce, 1 anise, 1T rock sugar, 1 slice of licorice root

Directions

1. Soak the dried shrimp in rice cooking wine for 10 minutes and remove. Press the water chestnuts, drain and minced. Tear the steam bun into small pieces and soak in a 1/4 cup of broth for about 5 minutes (Be careful not to overload the bum with broth. Just let it soak enough broth.) Cut the green onion in cross section.
2. Mince half of the dried shrimp and put in a bowl. Add the ground pork, minced green onions and tender ginger, then the whisked egg, seasonings, the moist steam bun, water chestnuts and tapioca starch. Mix well until the meat mixture becomes likes a paste.
3. Heat the pan. Add 1T of oil. First, add the remaining dried shrimp and stir-fry until fragrant. Next add the green onions sections and stir-fry until fragrant. Immediately add the Chinese cabbages and stir-fry briefly. Add 2T of rice cooking wine and cook for a while. Add the cooking sauce and the broth. Reduce the heat to low.
4. Make the meat mixture into balls. Place the meat balls in the direction 3. Put the lid on and simmer for 1 hour and 20 minutes. Serve.

Time-saving method

If you cook the meatballs in a Hotpan or Durotherm, just put the lid on and braise for 50 minutes.

～CC 烹調秘訣 | Cooking tips

1. 做法3.和4.在製作肉丸時，記得以小火煮，不可沸騰，以防止肉丸破碎。
2. 除了大白菜，也可用青江菜替代，另有一番滋味。而湯頭中如果多加入干貝及蛤蜊，風味更讚喔！

1. Remember to cook on low heat in direction no.3 and no.4 while making meatballs. Do not bring to a boil otherwise the meatballs will not be able to hold the shapes.
2. You can substitute bok choy for Chinese cabbage to bring out different flavor. If you add scallops and clams in the stock, the dish will be tastier.

東北燒肉

Northeastern-Style Stew Pork

這道東北燒肉是大陸東北地方的經典名菜，入口即化、口感層次分明是最大的特色。

This is dish is a classic cuisine in northeastern China. As you take a bit of it, not only the meat would just melt in your mouth but also you can taste the distinguish layers of textures.

材料

五花肉900克、南乳11/2大匙、蔥10支、薑5片、蒜仁1/2杯、甜麵醬31/2大匙、桂皮1支（6～8克）、米酒2大匙、老抽2大匙、醬油3大匙、水21/2杯

做法

1. 南乳搗碎；蒜仁整粒去皮。
2. 五花肉切長約4公分，放入鍋內，皮先朝鍋底煎，待煎至呈金黃色，倒入老抽拌炒均勻，取出。
3. 鍋燒熱，倒入1大匙油，放入整支蔥、薑、蒜炒至呈焦黃色，倒入甜麵醬，淋入米酒炒勻，續入五花肉，淋入醬油，倒入南乳，放入桂皮、水，蓋上鍋蓋煮，煮滾後改小火繼續煮約90分鐘。
4. 將做法3.倒入容器中，放入蒸籠（水滾了）蒸約30～40分鐘即可食用。

這樣做更省時

如果使用壓力鍋燉煮，做法3.不需加水，蓋上鍋蓋，加熱後待上升兩條紅線，再改小火煮約18分鐘。做法4.如果使用壓力鍋蒸，將食材倒入壓力鍋的深鍋，扣入壓力鍋內，壓力鍋內注入800c.c的水，待上升兩條紅線，再改小火煮約10分即可。

Ingredients

900g pork belly, 11/2T fermented red beancurd, 10 stalks of green onions, 5 slices of ginger, 1/2 cup of garlic cloves, 31/2T sweet soybean paste, 1 cinnamon stick (6g to 8g), 2T rice cooking wine, 2T dark soy sauce, 3T soy sauce, 21/2 cup water

Directions

1. Crush the fermented red beancurd. Peel the garlic cloves.
2. Cut the pork belly into 4cm wide sections. Add the pork, skin side down and sear until it turns golden. Add the dark soy sauce and stir evenly. Remove.
3. Heat the pan. Add 1T of oil. Stir-fry the green onion stalks, ginger and garlic until brown. Add the sweet soybean paste, pour in the rice cooking wine and stir evenly. Then add pork belly, soy sauce, fermented red beancurd, cinnamon stick and water. Cover and bring to a boil. Reduce the heat to low and simmer for 90 minutes.
4. Transfer the direction 3. into a bowl. Place it in a steamer over boiling water. Steam for 30 to 40 minutes and serve.

Time-saving method

If you are using a Duromatic to stew the food, then you don't need to add water in the direction 3. Just lock the lid and turn on the heat to high. Reduce to low heat when two red bars are visible and simmer for 18 minutes. To steam the food in a Duromatic, just put the ingredients in the deep pan and place in the pressure cooker. Add 800c.c. of water in the pressure cooker and turn on the heat. When two red bars are visible, cook the food for another 10 minutes on low heat.

CC 烹調秘訣 | Cooking tips

1. 老抽是指顏色深的醬油，呈有光澤的棕褐色，不具鹹味。一般是用在像紅燒時，替食物上色。
2. 南乳指紅色豆腐乳（紅腐乳、紅糟豆腐乳），帶有一點甜味，一般超市或雜貨店有賣。
3. 辛香料一定要爆炒至焦黃，煮好的整鍋肉才會香。

1. Dark soy sauce is a darker soy sauce with the color of glossy brown. It doesn't contain saltiness and is usually used to add color during stewing.
2. Fermented red beancurd is a kind of fermented beancurd in red color (like red preserved beancurd or fermented beancurd with red yeast rice). It offers a light, sweet flavor and can be found in supermarkets and grocery stores.
3. All spices need to be stir-fry until fragrant to bring out the aromatics of the meat.

彩頭南乳蒜燒肉

Braised Pork with Chinese Radish,
Fermented Red Beancurd and Garlic

如果你是不喜歡吃五花肉的皮和肥肉的人，那
一定不能錯過這道不肥膩的肉料理，它一定讓
你對五花肉改觀而愛上它。配料的白蘿蔔鮮甜
軟透，受歡迎程度更不輸滷肉。

If you are not in falvor of the skin and fat of pork belly,
don't miss this ungreasy meat dish. It will change your
perspective in pork belly and make you fall in love with
it. Furthermore, because the daikon tastes tender and
delicious, it is as popular as the braised pork.

材料

五花肉1,000克、白蘿蔔1根、蒜苗3支、蒜仁1杯、八角2粒、紅辣椒1支、紹興酒1/2杯、水1杯

調味料

南乳11/2塊、南乳汁1大匙、醬油2大匙、醬油膏1大匙、老抽2大匙

做法

1. 五花肉切長3～4公分，皮先朝鍋底煎，待煎至呈金黃色取出。

2. 白蘿蔔切圓厚片；1支蒜苗和紅辣椒切絲，浸泡冷水約6分鐘；2支蒜苗切段；南乳搗碎。

3. 鍋燒熱，倒入11/2大匙油，放入蒜苗、1/2杯的蒜仁爆炒至金黃色，倒入五花肉，淋入紹興酒，待酒精揮發後倒入調味料、八角、白蘿蔔和水開始煮，煮滾後再煮50～60分鐘，最後倒入剩餘的蒜粒煮約5分鐘，盛盤，裝飾少許蒜苗絲和紅辣椒絲即可食用。

這樣做更省時

如果使用壓力鍋燉煮則不需加水，加熱後上升兩條紅線改小火16分鐘；使用双享鍋也不需加水，煮約30分鐘。

Ingredients

1,000g pork belly, 1 chinese radish, 3 stalks of leek, 1 cup of garlic cloves, 2 anise, 1 red chillies, 1/2 cup of Shaohsing wine, 1 cup of water

Seasoning

11/2 squares of fermented red beancurd, 1T fermented red beancurd juice, 2T soy sauce, 1T soy sauce paste, 2T dark soy sauce

Directions

1. Cut the pork belly into 3 to 4cm-wide strips. Sear in the pan with skin side down until golden brown. Remove.
2. Cut the chinese radish into thick rounds. Shred 1 stalk of leek and red chillies soak in cold water for 6 minutes, Cut 2 stalks of leek into sections. Mash the fermented red beancurd.
3. Heat the pan. Add 11/2 T of oil. Stir-fry leeks and 1/2 cup of garlic cloves until golden. Add the pork belly and Shaohsing wine. Cook until the alcohol evaporates. Add the seasonings, anise, chinese radish and water and bring to a boil. After cooking for 50 to 60 minutes, add the remaining garlic cloves and cook for another 5 minutes. Transfer to a serving plate. Sprinkle with shredded garlic and red chillies serve.

Time-saving method

When cooking in a Duromatic, you don't need to water. Reduce to low heat when two red bars are visible and simmer for 16 mimutes. The same rule applies to Durotherm and cook for 30 minutes.

CC烹調秘訣 | Cooking tips

1. 酒精一定要煮至揮發，例如加入2大匙的酒煮至剩1大匙，就表示酒精揮發了，如果沒有揮發的話會有酸澀味。
2. 蒜苗絲浸泡冷水，吃的時候才會脆。
3. 滷肉時不加入水滷，滷出來的肉才會香，但一般鍋子都需要加入水，只有鎖水性強的鍋子才可以不加水滷肉，建議可到市面上尋找這類鍋子使用。
4. 在做法3.中放入蒜仁，整鍋肉會更香，蒜仁也會變得更好吃。

1. When cooking with wine, the alcohol needs to be evaporated thoroughly. For example, if you add 2T of wine in the cooking process, when there is only 1T of wine left in the pan, it means that the alcohol has been evaporated. Otherwise the meat will taste bitter and sour.
2. Soaking shredded leek in cold water can keep the crunchy texture.
3. Braising meat without adding any water can make the aroma stronger. However, you usually have to add water when using ordinary pans. You can only braise meat with no water in pans with a strong water-locking characteristic. Look for this kind of pans sold on the market.
4. Add garlic cloves in the direction No.3 makes the aroma of the whole dish stronger. The garlic itself tastes better, too.

台式紅糟肉

Taiwanese Red Vinasse Pork Belly

每次到黑白切麵店或粥店，只要看到紅糟肉一定會點來品嘗。不過市售的紅糟肉價格不便宜，不妨試試在家自己做做看！

Every time I go to a noodle or congee shop, I always order a dish of red vinasse pork belly if it's on the menu. But the price is always not cheap. So why not try to make it at home!

材料

豬五花肉（松阪豬、小里肌或胛心肉都可以）600克、雞蛋1個、地瓜粉1/3杯

醃肉料

紅糟2 1/2大匙、蒜末1大匙、香蒜粉1/2小匙、醬油1大匙、細砂糖1/2大匙、米酒2 1/2大匙、白胡椒粉適量、太白粉1/2大匙

做法

1. 豬肉放入醃肉料中醃約2小時。
2. 雞蛋打散拌勻，將蛋液倒入做法1.中拌勻。
3. 將豬肉沾裹地瓜粉，放入油溫160℃的油鍋中炸至呈金黃色，取出瀝乾油分。
4. 將炸好的豬肉切片盛盤，可搭配涼拌小黃瓜更美味。

Ingredients

600g pork belly (Matsuzaka pork, pork tenderloin or pork shoulder is fine, too), 1 egg, 1/3 cup of sweet potato flour

Marinade

2 1/2T red vinasse, 1T minced garlic, 1/2t garlic powder, 1T soy sauce, 1/2T granulated sugar, 2 1/2T rice cooking wine, white pepper as need, 1/2T tapioca starch

Directions

1. Marinate the pork for 2 hours.
2. Whisk the egg. Add the whisked egg into direction 1. and mix thoroughly.
3. Coat the pork with sweet potato flour. Deep-fry in the oil at 160°C until golden brown. Remove and Drain.
4. Slice the cooked pork and arrange on the plate. It tastes even better with cucumber salad.

CC 烹調秘訣 | Cooking tips

1. 當肉放入油鍋中炸時，不要馬上翻動，以防止皮剝落。此外，不要持續用中大火炸，肉放入鍋後用中小火，要起鍋前再用中大火炸。
2. 涼拌小黃瓜DIY：準備小黃瓜300克、紅辣椒2支、蒜末2大匙、鹽2/3大匙和調味料（細砂糖2大匙、白醋4大匙、香油1大匙）。小黃瓜切圓片，放入些許鹽拌勻，靜置15分鐘後瀝乾水分，放入容器中。加入調味料、蒜末和紅辣椒段拌勻，放入冰箱冷藏至少1小時，使其入味即可。
3. 有些店家因為加入了色素，賣的紅糟肉會比較紅，建議自己做更營養健康。

1. Deep-fry the meat in the fry-pan. Don't flip the meat right away. Otherwise the skin will fall off. Don't keep deep-frying over medium-high heat. Turn the heat to medium-low when placing the meat into the pan and increase to medium-high heat before removing from the pan.
2. **How to make cucumber salad:** prepare 300g cucumber, 2 red chillies, 2T minced garlic, 2/3T salt and seasonings (2T granulated sugar, 4T white vinegar, 1T sesame oil). Slice cucumbers into rounds and mix with salt thoroughly. Set aside for 15 minutes and drain. Put in a bowl. Stir in seasonings, minced garlic and chopped red chillies thoroughly. Put in the refrigerator for at least an hour to make it tastier.
3. The color of red vinasse pork belly sold in some shops is sometimes redder. That's because they add artificial colors in it. Homemade red vinasse pork belly is healthier and more nutritious.

芋香酥拌肉

Shredded Pork with Crispy Shredded Taro

芋香肉包裹著清脆的生菜葉一起吃，口感爽脆之外，我發現香酥的芋絲和軟嫩的肉絲真是絕配。

Wrapping the shredded pork and crispy taro with crunchy lettuce leave provides a refreshing texture. I also notice that crispy shredded taro goes perfectly well with tender shredded pork.

材料

芋頭300克、小里肌200克、蒜末1大匙、蔥末2大匙、紅辣椒末1 1/2大匙、香菜末2大匙、生菜150克

醃肉料

醬油1大匙、白砂糖少許、白胡椒粉適量、香油1/2大匙、米酒1/2大匙、蛋白1/3個、太白粉1 1/2大匙

調味料

白胡椒粉適量、鹽少許

做法

1. 芋頭削除外皮後切絲；小里肌切絲，放入醃肉料中醃約30分鐘。
2. 將芋頭放入油溫160℃的油鍋中炸至香酥，撈出瀝乾油分，再放入醃好的肉絲，炸至肉絲略呈白色，撈出瀝乾油分。
3. 鍋燒熱，倒入1大匙油，依序放入蒜末、蔥末和紅辣椒末爆香，續入炸過的肉絲拌炒幾下，再放入炸過的芋絲，撒上白胡椒粉、鹽、香菜末，拌炒幾下即可。
4. 包上生菜一起食用。

Ingredients

300g taro, 200g pork tenderloin, 1T minced garlic, 2T minced green onions, 1 1/2T minced red red chillies, 2T minced coriander, 150g lettuces

Marinade

1T soy sauce, granulated sugar to taste, white pepper as needed, 1/2T sesame oil, 1/2T rice cooking wine, 1/3 egg white, 1 1/2T tapioca starch

Seasonings

White pepper as needed, salt to taste

Directions

1. Peel the taro and shred. Shred the pork tenderloin and marinate for 30 minutes.
2. Deep-fry the shredded taro in oil at 160°C until crispy. Remove and drain. Deep-fry the marinated pork until it turns whitish. Remove and drain.
3. Heat the pan. Add 1T of oil. Add minced garlic, green onions and red chillies in order and stir-fry until fragrant. Add the deep-fried shredded pork and stir-fry for a little bit. Add the deep-fried shredded taro and sprinkle with white pepper, salt, minced coriander. Stir-fry briefly.
4. Wrap with lettuces and serve.

 CC 烹調秘訣 | Cooking tips

1. 生菜可選用美生菜或萵苣（俗稱廣東A），但記得洗淨後要先放入冰水中冰鎮約15分鐘，瀝乾水分後吃才會清脆。
2. 做法3.的拌炒時間不宜太久。

1. When selecting lettuces, you can choose iceberg lettuce or romaine lettuce. In order to keep the crunchy texture, remember to soak in ice water for 15 minutes after washing. Then drain and serve.
2. When stir-frying the ingredients in direction No.3, don't stir-fry too long.

蛋黃蒸肉

Steamed Ground Pork with Salted Egg Yolk

重口味的鹹蛋黃和醃漬食材，純樸的風味令我想起孩提時代的餐桌，有了它這一餐一定吃很飽，是很下飯的一道料理喔！

The salted egg york and the preserved ingredients are both strongly flavored. This simple and humble dish brings me back the memories of childhood. If this dish was on the dining table, I knew I would be very full at that meal. It went very well with rice!

材料

絞肉600克、鹹蛋黃3個、罐頭蔭瓜末1/4杯

調味料

白胡椒粉1/3小匙、細砂糖1小匙、醬油2大匙、醬油膏1大匙

做法

1. 將除了鹹蛋黃以外的材料和調味料拌勻成肉餡。
2. 將鹹蛋黃塗抹上一層米酒，置於一旁10分鐘後對切。
3. 將肉餡放入可蒸煮的器皿內，鋪上鹹蛋黃，放入蒸籠（水滾了）蒸50分鐘，取出即可食用。

這樣做更省時

如果使用壓力鍋燉煮，將做法**3.**放入壓力鍋的淺蒸盤內，倒入800c.c的水，將淺蒸盤扣住蓋上鍋蓋，加熱後上升成兩條紅線時，改小火15分鐘。

Ingredients

600g ground pork, 3 salted egg yolk, 1/4 cup of minced canned pickled cucumbers

Seasonings

1/3 t white pepper, 1t granulated sugar, 2T soy sauce, 1T soy sauce paste

Directions

1. Mix all the ingredients and seasonings into meat mixture, except the salted egg yolk.
2. Brush a layer of rice cooking wine on the salted egg yolk. Set aside for 10 minutes and cut in half.
3. Place the meat mixture in a steamable pan. Lay the salted egg yolk. Place in a steamer with boiling water for 50 minutes. Remove and serve.

Time-saving method

When cooking with a Duromatic, place the meat mixture in direction **3.** in the pressure cooker's shallow pan. Add 800c.c. of water. Secure the shallow pan. Lock the lid and turn on the heat. When two red bars are visible, reduce to low heat for 15 minutes.

CC 烹調秘訣 | Cooking tips

如果再加入少許冬菜末、紅辣椒末，就成了重口味愛好者的最愛。

If you crave for a stronger taste, just add a pinch of minced preserved cabbage and red chillies.

客家小炒
Hakka Stir-Fried Dish

客家小炒要怎麼烹調才夠味呢？這道菜要好吃，刀工和料理時間都很重要，只要有耐心掌握火候就容易成功！

How to cook a tasty Hakka stir-fried dish? The key points are the cutting skills and cooking time. As long as you patiently control the cooking temperature, you will be able to make this dish successfully.

材料

乾魷魚120克、五花肉150克、豆乾120克、蔥1支、芹菜4支、紅辣椒2支、蒜苗2支、乾香菇6朵、蒜仁3粒

調味料

米酒1大匙、醬油2 1/2大匙、糖1/3大匙、白胡椒粉少許、沙茶醬1/3大匙

做法

1. 乾魷魚泡水約30分鐘,取出切條狀;五花肉和豆乾都切條狀;蔥和芹菜都切段;紅辣椒和蒜苗都切斜段;乾香菇泡水至軟,取出切條狀;蒜仁切片。
2. 鍋燒熱,倒入2大匙油,先放入五花肉爆炒至焦黃色,取出。然後放入豆乾爆炒至呈金黃色,取出。
3. 原鍋放入蒜仁、蔥段、紅辣椒爆香,續入香菇、魷魚炒約2分鐘,再放入五花肉、豆乾炒勻,加入調味料略炒,最後倒入芹菜、蒜苗快炒幾下即可食用。

Ingredients

120g dried squid, 150g pork belly, 120g dried tofu, 1 stalk of green onions, 4 stalks of celery, 2 red chillies, 2 stalk of leek, 6 dried shitake mushrooms, 3 garlic cloves

Seasonings

1T rice cooking wine, 2 1/2T soy sauce, 1/3T sugar, white pepper to taste, 1/3T Shacha sauce

Directions

1. Soak the dried squid in water for 30 minutes. Remove and cut into strips. Cut the pork belly and dried tofu into strips. Cut the green onions and celery into sections. Cut the red chillies and leeks in cross section. Soak the dried shitake mushrooms until softened, remove and cut into strips. Slice the garlic clove.
2. Heat the pan. Add 2T oil. Stir-fry the pork belly first until brown. Remove. Stir-fry the dried tofu until golden brown. Remove.
3. In the same pan. Stir-fry the sliced garlic, green onions sections and red chillies until fragrant. Then add the shitake mushrooms and squid. Stir-fry for 2 minutes. Stir in the pork belly and dried tofu. Add the seasonings and stir slightly. Last, add celery and leek and stir-fry quickly. Serve.

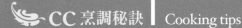

CC 烹調秘訣 | Cooking tips

1. 如果不喜歡吃辣,可在起鍋前再放入紅辣椒即可。
2. 五花肉、豆乾一定要爆炒得呈金黃色才行,這樣才能炒出香氣、口感佳。
3. 這道菜中沙茶醬不要加太多,只要些許沙茶醬就能增加風味。

1. If you don't like spicy food, just add red chillies right before removing food from the pan.
2. Pork belly and dried tofu must be stir-fried until golden brown to bring out the aroma and texture.
3. Don't add too much Shacha sauce in this dish. Just a little bit to add a kick of flavor.

客家高麗菜封肉

Hakka-style Braised Pork Belly
with Taiwanese Cabbages

這道客家美食的特色，在於燉煮過程中，將高
麗菜的鮮甜味和肉融合，也使湯汁更加鮮美，
吃過的人都回味再三。

The characteristic of this Hakka dish is that during the
process of braising, the freshness of Taiwanese cabbages
not only blends into the meat but also enhances the
deliciousness of the soup. It's a taste that will linger in
your mouth and makes you crave for more.

材料

五花肉900克、高麗菜1顆（約600克）、去皮蒜仁8粒、蔥6支、高湯11/2杯、醃冬瓜3大匙、豆醬和汁液共2大匙、醬油1/2杯、香菜適量、嫩薑片3片

醃肉料

蔥2支、薑片3片、米酒2大匙、醬油3大匙

做法

1. 醃肉料中的蔥和薑片拍扁；五花肉切4公分寬，放入醃肉料中拌勻，醃約30分鐘；高麗菜以十字切法將芯取出；高湯做法參照p.117。
2. 鍋燒熱，倒入1大匙油，放入五花肉煎至呈金黃色，取出。也可放入上下火220℃的烤箱中烤至呈金黃色，但仍需視自家烤箱斟酌的時間。
3. 取一湯鍋，鍋燒熱，倒入1大匙油，先放入蔥、嫩薑片和蒜仁爆香，續入整顆取出芯的高麗菜，將五花肉放入高麗菜的中間，淋上醬油，倒入高湯，放入醃冬瓜、豆醬，燉煮至高麗菜變軟，盛盤後撒上香菜即可食用。

Ingredients

900g pork belly, 1 Taiwanese cabbages (about 600g), 8 peeled garlic cloves, 6 stalks of green onions, 1½ cup of broth, 3T pickled white gourds, 2T salted bean paste and juice, 1/2 cup of soy sauce, coriander as needed, 3 slices of tender ginger

Marinade

2 stalks of green onions, 3 slices of ginger, 2T rice cooking wine, 3T soy sauce

Directions

1. Press the green onions and sliced ginger used in the marinade. Cut the pork belly into 4cm thick pieces. Marinate for 30 minutes. Cross cut the bottom of cabbage to remove the central core. See p.117 for the broth recipe.
2. Heat the pan. Add 1T oil. Sear the pork belly until golden brown. Remove. Or you can bake it in the oven at 220℃ until golden brown. The baking time may vary, depending upon the oven.
3. Heat a stock pot. Add 1T oil. Stir-fry green onions, tender ginger and garlic first until fragrant. Add the whole Taiwanese cabbages with central core removed. Place the pork belly in the middle of the cabbages. Add the soy sauce, broth, pickled white gourds and salted bean paste. Cook until the cabbages are softened. Transfer to a serving plate. Sprinkle with coriander and serve.

 CC 烹調秘訣 | Cooking tips

1. 醃冬瓜罐中就會有豆醬和汁液。自己醃的冬瓜和市售醃冬瓜都可以做這道料理，但如果買不到醃冬瓜，可用蔭瓜代替。
2. 醃冬瓜DIY：將1,000克的冬瓜削除外皮後切3公分的塊狀，放入鋼盆中，均勻地撒入80克的鹽，醃漬一晚讓冬瓜逼出苦水，瀝乾水分。然後將80克豆粕、2大匙鹽和1小匙咖啡用紅糖拌勻成拌料。接著將冬瓜排入容器中，均勻地撒上拌料，重複一層冬瓜一層拌料的方式疊放至材料完全放入，加入4片甘草、50c.c.的米酒後密封，放在陰涼處醃2個月即可。

1. You can find salted bean paste and juice in a pickled white gourd jar. You can use either homemade pickled white gourds or the ones sold in stores to make this dish. If you cannot find pickled white gourds, substitute with pickled cucumbers.
2. How to make pickled white gourds: peel 1000g of white gourd and cut into 3cm-thick pieces. Put in a mixing bowl. Sprinkle evenly with 80g of salt. Set aside for one night to draw out excess bitter water. Drain. Stir 80g of soybean meal, 2T salt and 1t coffee with brown sugar for seasoning. Arrange a layer of white gourds in a container. Evenly sprinkle with the seasonings. Then repeat the process until all the ingredients are added in the container. Add 4 slices of licorice root and 50c.c. of rice cooking wine. Seal tightly. Preserve for 2 months in a cooling area.

Q彈豬腳花生

Braised Pork Knuckles with Peanuts

油而不膩、肉質軟爛、皮帶彈性的豬
腳，一口咬下的滿足感，是老饕口中的
終極美味。

It's fatty but not greasy and has tender meat
with chewy skins. The complete satisfaction
that every bite brings makes this dish the
ultimate savor among epicures.

材料

黑毛豬腳1隻（約1,500克）、生花生300克、蔥12支、蒜仁1/2杯、紹興酒11/2杯、八角2粒、花椒粒1/2大匙、冰糖21/2大匙、水2杯

調味料

醬油41/2大匙、醬油膏41/2大匙、老抽4大匙

做法

1. 豬腳剁塊狀；生花生放入水中泡約2～3小時。
2. 取一深鍋，鍋燒熱，倒入1大匙油，放入整支蔥和蒜仁爆香至焦黃。
3. 放入豬腳，淋入紹興酒，煮至酒精揮發，放入生花生、八角和花椒粒，加入調味料，放入冰糖，然後倒入水煮約1小時30分鐘即可食用。

這樣做更省時

燉煮豬腳比較費時，如果使用壓力鍋燉煮，加熱後等上升兩條紅線，再改小火煮約18分鐘即可。此外，還可以不加3杯水，調味料刪減成醬油31/2大匙、醬油膏31/2大匙、老抽3大匙就足夠了。

Ingredients

1 black pig knuckle (about 1,500g), 300g raw peanuts, 12 stalks of green onions, 1/2 cup of garlic cloves, 11/2 cup of Shaohsing wine, 2 anise, 1/2T Sichuan peppercorns, 21/2 T rock sugar, 2 cup of water

Seasonings

41/2 T soy sauce, 41/2 T soy sauce paste, 4T dark soy sauce

Directions

1. Chop the pig knuckle into pieces. Soak the raw peanuts in water for 2 to 3 hours.
2. Take a deep pan. Heat the pan and add 1T of oil. Stir-fry the whole stalks of green onions and garlic cloves until brown.
3. Add the pig knuckle. Add the Shaohsing wine and cook until the alcohol evaporates. Next, add the raw peanuts, anises and Sichuan peppercorns. Then add the seasonings and rock sugar. Last, add the water and cook for one and half hour. Serve.

Time-saving method

It takes longer time to braise pig knuckles, but using a Duromatic to do the job can save your time. After turning on the heat, just wait until two red bars are visible and then reduce to low heat for 18 minutes. Furthermore, you don't need to add water. The amount of seasonings needs be reduced to 31/2 T soy sauce, 31/2 T soy sauce paste, 3T dark soy sauce.

CC 烹調秘訣 | Cooking tips

1. 蔥、蒜一定要爆香至焦黃才可以，這動作是製作這道菜的關鍵，這一鍋豬腳的香氣就靠這個步驟了。
2. 紹興酒一定要煮到酒精揮發，否則會酸澀。紹興酒比米酒香很多，做這道料理再適合不過了。
3. 八角和花椒粒放入小白布袋中，綁緊再放入鍋中，烹調時不會四處亂散，方便食用。

1. The key of making this dish is to stir-frying the green onions and garlics until brown. It will bring out the aroma of braised pig knuckles.
2. The alcohol of Shaohsing wine must be cooked until evaporates. Otherwise it will taste bitter and sour. The aroma of Shaohsing wine is much stronger than that of rice cooking wine, making the wine an ideal ingredient for that dish.
3. You can put the anises and peppercorns in a small white cloth bag. Tie it before putting in the pan. This can prevent the spices floating everywhere in the pan and make it easier to savor the dish.

津白牛肉煲

Beef with Chinese Cabbage Casserole

津白就是大白菜，以它釋放出的鮮甜來烹調的料理，是調味料比不上的。我推薦冬天一定要吃，大大補充元氣。

This is a dish cooked with the delicious and fresh flavor of Chinese cabbage. It's something that no seasonings can compare with. It's a must dish in cold winter to revive vitality.

材料

牛肋條（牛腩）600克、大白菜400克、蔥末2大匙、薑末1 1/2大匙、蒜末1大匙、紅辣椒末1大匙、洋蔥1/2個、高湯1,500c.c.、蔥1支、薑片3片、冬粉2把、香菜適量

調味料

辣豆瓣醬2大匙、米酒2大匙、冰糖1大匙、醬油3大匙、八角2粒

做法

1. 牛肋條切塊狀；高湯做法參照p.117；洋蔥切絲；蔥切段；冬粉浸泡水至軟，再剪成適當的長度。
2. 鍋燒熱，倒入1 1/2大匙油，先放入牛肋條煎至呈金黃色，取出。
3. 原鍋放入蔥末、薑末、蒜末和紅辣椒末爆香，續入洋蔥絲，以小火炒約2分鐘，加入辣豆瓣醬炒至散發出香氣，淋入1大匙米酒炒勻，再倒入冰糖、醬油、八角、高湯（先留3大匙不要倒入）和牛肋條煮約1小時30分鐘。
4. 另一鍋燒熱，倒入1大匙油，先放入蔥段、薑片爆香，續入大白菜，淋入剩餘的米酒和3大匙的高湯，煮至大白菜變軟，放入冬粉和做法3.，最後撒上香菜即可食用。

這樣做更省時

使用壓力鍋燉煮牛肋條，只需加入800c.c.的高湯，減少1/4量的醬油，加熱後等上升兩條紅線，再改小火煮約18分鐘即可。

Ingredients

600g short rib(or beef shank), 400g Chinese cabbages, 2T minced green onions, 1 1/2T minced ginger, 1T minced garlic, 1T minced red chillies, 1/2 onion, 1,500c.c. broth, 1 stalk of green onion, 3 slices of ginger, 2 bunches of green bean noodles, coriander as needed

Seasonings

2T spicy soy bean paste, 2T rice cooking wine, 1T rock sugar, 3T soy sauce, 2 anise

Directions

1. Cut the short rib into bite-size pieces. See p.117 for the broth recipe. Shred onion. Cut the green onions into sections. Soak the green bean noodles until softened and cut into proper lengths.
2. Heat the pan. Add 1 1/2T of oil. Sear the short rib until golden brown. Remove.
3. In the same pan. Stir-fry the minced green onions, ginger, garlic, red chillies until fragrant. Add shredded onions and stir-fry for 2 minutes on low heat. Stir in the spicy soy bean paste until fragrant. Add 1T rice cooking wine and mix well. Then add rock sugar, soy sauce, anise, broth (save 3T for later use) and short rib. Cook for 1 hour and 30 minutes.
4. Heat another pan. Add 1T oil. First, stir-fry the green onion sections and sliced ginger until fragrant. Add the cabbage, the remaining rice cooking wine and the 3T broth. Cook until the cabbages are tender. Add the green bean noodles and the direction **3**. Last, sprinkle with coriander and serve.

Time-saving method

When using a Duromatic to stew beef short rib, you only need to add 800c.c. of broth and reduce 1/4 amount of soy sauce. When the two bars are visible, reduce to low heat and cook for 18 minutes.

CC 烹調秘訣 Cooking tips

可用豬腱肉取代牛肋條，一樣美味。
You can substitute pork shank for beef short rib. It tastes delicious, too.

Rice, Noodles and Congee Section

Part 4 飯、麵、粥類

肚子餓的時候，會想吃什麼？
吃碗飯或麵，或是來個粥也挺不錯。
除了滿滿幸福的飽足感，更會讓人充滿活力。
所以一日三餐之中，
怎麼能少了元氣補充聖品的飯、麵、粥呢？
快來一碗散發誘人香氣，
引人食指大動的美味主食吧！

What are you craving for when you are hungry?
A bowl of rice, noodles or congee sounds like a good choice!
It not only gives you a full feeling of happiness but also boosts the vitality.
It's a must-have energy booster food in our three meals of a day.
So hurry to have a bowl of rice,
noodles or congee with indulging aroma and flavor.

嘉義雞肉飯

Chiayi Chicken Rice

雞油加上油蔥酥的香氣，讓嘉義雞
肉飯成了遠近馳名的美食，是到嘉
義不可錯過的小吃。不過，也可以
自己在家做喔！現在就來試試。

The aroma of deep-fried shallots with
chicken fat makes Chiayi chicken rice
famous for miles around. It's a dish that
you cannot miss when visiting Chiayi.
However, you can also make it at home,
too! Let's give it a try.

材料

去皮去骨雞胸肉300克、紅蔥頭末150克、雞油300克、白飯適量、香菜末適量

醬汁

醬油1/4杯、水3大匙、冰糖1/2大匙

煮料

米酒1大匙、紅蔥頭3粒、五香粉1/4小匙

做法

1. 取3粒紅蔥頭拍扁；醬汁的材料煮滾。
2. 將煮料倒入鍋中，放入雞胸肉，蓋上鍋蓋，燜煮至雞胸肉熟，取出剝成絲狀。
3. 將雞油放入鍋內，炸乾雞油呈褐色（先中火再改小火），再撈出雞油渣，留下金黃色的雞油，放入紅蔥頭末，炸至酥脆且呈金黃色，取出。
4. 取適量炸好的紅蔥頭酥（油蔥酥）放入醬汁中，再加入1大匙雞油渣拌勻。
5. 將白飯盛入碗中，鋪上雞絲，淋上做法**4.**，撒上少許香菜末即可食用。

這樣做更省時

烹煮雞胸肉時，如果使用休閒鍋或双享鍋，蓋上鍋蓋加熱，待冒煙後改成小火煮4分鐘，移至外鍋燜8分鐘即可。

Ingredients

300g skinless and boneless chicken breast, 150g minced shallots, 300g chicken fat, rice as needed, minced coriander as needed

Sauce

1/4 cup of soy sauce, 3T water, 1/2 T rock sugar

Seasonings for cooking chicken

1T cooking wine, 3 shallots, 1/4t five spices powder

Directions

1. Press 3 shallots. Bring the sauce to a boil.
2. Add the seasonings for cooking chicken in a pan. Add the chicken and put the lid on. Simmer until the chicken is cooked thoroughly. Remove and tear into shreds.
3. Add the chicken fat in a pan. Deep-fried the chicken fat until it turns brown (first on medium heat and then reduce to low heat). Skim the scum and keep the golden chicken oil, add the minced shallots , fried till gold and crumbly,remove.
4. Stuff enough amount of deep-fried shallots into the sauce,and add 1 T chicken fat and mix well.
5. Fill a bowl with rice. Top with chicken shreds. Drizzle with direction **4.** Sprinkle with a pinch of minced coriander and serve.

Time-saving method

When cooking chicken breast in a Hotpan or Durotherm, you need to cover and turn on the heat. Reduce to low heat when the steam starts to hiss out of the pan and cook for 4 minutes. Transfer to outer pot and set aside for 8 minutes. Serve.

 CC 烹調秘訣 | Cooking tips

1. 紅蔥頭酥（油蔥酥）DIY：將紅蔥頭末、片或細絲放入160℃的油鍋中炸，發現紅蔥頭即將變成金黃色時要趕緊準備撈起，不要等到已經都變成金黃色才要撈出，錯過時機紅蔥頭就會變黑且發出苦味。
2. 雞油可到傳統菜市場中向雞販購買即可。
3. 炸好的紅蔥頭可以用來拌燙青菜或是加入湯麵中。

雞油

1. **How to make deep-fried shallots at home:** Deep-fry minced, sliced or shredded red onions in oil at 160℃. Remove the red onions immediately when it starts to turn golden brown. Don't wait until shallot is already golden brown otherwise it will turn black and taste bitter.
2. Chicken fat can be bought at chicken vendors in traditional markets.
3. Deep-fried shallots can be mixed with blanched vegetables or soup noodles.

香菇肉燥飯

Rice with Shitake Mushroom
Flavored Pork

就是滷肉飯，真是令人懷念的古早
味啊！我喜歡一次製作大鍋的肉
料，拿來拌飯、拌麵、拌青菜，非
常實用，可以說是萬用醬料。

It is also known as rice with braised
ground pork, a very traditional dish worth
remembering. I like to cook a big pot of
braised ground pork every time. It's a very
useful "whole-purpose" sauce. You can mix
it with rice, noodles and vegetables.

材料

五花肉900克、乾香菇12朵、紅蔥頭末1/4杯、油蔥酥1/4杯、香蒜酥1/4杯、甘蔗頭1個或黑糖1大匙、白飯適量、香菜少許、水2杯

調味料

醬油1/4杯、醬油膏1/4杯、老抽2大匙、八角1粒、五香粉1/2小匙、水1杯

做法

1. 五花肉切小丁；乾香菇泡水至軟，然後切小丁。
2. 鍋燒熱，倒入2大匙油，放入五花肉爆炒至呈金黃色，先取出。放入紅蔥頭末炒至焦黃香酥，再倒入香菇炒香，倒回五花肉一起拌炒。
3. 繼續倒入調味料，加入甘蔗頭或黑糖以及水煮約30分鐘，再加入油蔥酥、香蒜酥煮約50分鐘，即成肉燥料。
4. 將白飯盛入碗中，淋上肉燥料，撒入香菜即可食用。

這樣做更省時

如果使用壓力鍋製作肉燥料，只要倒入1/4杯水，先放入做好的做法**2.**，再放入其他所有材料，蓋上鍋蓋，加熱後等上升兩條紅線，再改小火煮約18分鐘即可。

Ingredients
900g pork belly, 12 dried shitake mushrooms, 1/4 cup minced shallots, 1/4 cup of crisp red onions, 1/4 cup minced deep-fried garlics, 1 piece of sugar can root or 1T brown sugar, rice as needed, a pinch of corianders, 2 cups of water

Seasonings
1/4 cup of soy sauce, 1/4 cup of soy sauce paste, 2T dark soy sauce, 1 anise, 1/2t five spices powder, 1 cup of water

Directions
1. Dice the pork belly. Soak the dried shitake mushrooms until softened and dice into small pieces.
2. Heat the pan. Add 2T of oil. Stir-fry the pork belly until golden brown. Remove. Stir-fry the minced shallots until crispy, brown and fragrant, and add the shitake mushrooms. Stir-fry until fragrant. Add the pork belly and stir evenly.
3. Add the seasonings, sugar can root or brown sugar and water. Cook for 30 minutes. Stir in the deep-fried shallots and deep-fried garlics. Cook for 50 minutes. Then the meat sauce is ready to serve.
4. Fill a bowl with rice. Top with the meat sauce. Sprinkle with coriander. Serve.

Time-saving method
When cooking meat sauce in a Duromatic, you need to add 1/4 cup of water, then the meat sauce in direction **2.** (which is already done) and the remaining ingredients. Put the lid on and turn on the heat. When two red bars are visible, reduce to low heat for 18 minutes.

CC 烹調秘訣 | Cooking tips

1. 可用絞肉取代五花肉，多加了油蔥酥會更香。
2. 肉燥料可以一次製作大量再分成數包，放入冰箱以冷凍保存，食用時再解凍加熱，淋在燙熟的空心菜、地瓜葉、A菜上，或者拌麵、夾饅頭，既方便又好吃。

1. You can substitute ground pork for pork belly. Adding deep-fried shallots can enhance the flavor.
2. You can cook a lot of meat sauce at one time. Then divide it to several bag and store in the freezer. Defrost and heat it before serving. You can spoon it on blanched Chinese water spinach, sweet potato leaves or Arden lettuce. It also goes well with noodles and steam buns. It's very yummy and convenient.

焢肉飯

Rice with Braised Pork

焢肉飯是台灣最具代表性的小吃之一，以白飯搭配滷得滑嫩香Q的肉塊，淋上香噴噴的滷肉汁，若再來幾盤小菜，就是豐盛的一餐。

Rice with braised pork is one of the most representative street snacks in Taiwan. After topping the rice with tasty and tender meat and drizzle with aromatic sauce, just add a couple more side dishes, then you will have a substantial meal.

材料

五花肉900克、蔥6支、去皮蒜仁70克、八角2粒、五香粉1/2小匙、水1杯、白飯適量、紹興酒1杯

調味料

醬油1/4杯、醬油膏1/4杯、冰糖2大匙、老抽3大匙

做法

1. 五花肉切成厚1.3～1.5公分，長10～12公分。
2. 鍋燒熱，倒入1 1/2大匙油，放入五花肉，皮先朝鍋底煎，待煎至呈金黃色取出。
3. 然後放入蒜仁、整支蔥爆炒至焦黃，放入煎好的肉，淋入紹興酒，待酒精揮發，倒入調味料和水、八角、五香粉，燉煮約1小時10分鐘，時間可視個人喜愛的肉Q軟度來決定。
4. 將白飯盛入碗中，放上焢肉，淋上滷肉汁即可食用。

這樣做更省時

如果使用壓力鍋燉煮滷肉，不需放水，醬油減量，直接加熱，待上升兩條紅線改小火煮約16分鐘；如果使用双享鍋燉煮，不需放水煮約45分鐘；如果使用休閒鍋，需加入1/4杯水煮約25分鐘，再移入外鍋燜30分鐘即可。

Ingredients

900g pork belly, 6 stalks of green onions, 70g peeled garlic cloves, 2 anise, 1/2t five spices powder, 1 cup of water, rice as needed, 1 cup Shaohsing wine

Seasonings

1/4 cup of soy sauce, 1/4 cup of soy sauce paste, 2T rock sugar, 3T dark soy sauce

Directions

1. Cut the pork belly into 1.3 to 1.5cm thick, 10 to 12cm wide pieces.
2. Heat the pan. Add 1 1/2 T of oil. Sear the pork belly with the skin side down until golden brown. Remove.
3. Stir-fry the garlic cloves and the green onions stalks until brown. Add the seared pork, pour in Shaohsing wine and cook until the alcohol evaporates thoroughly. Add the seasonings, water, anise, five spices powder and simmer for 1 hour and 10 minutes, depending on the level of tenderness that your prefer.
4. Fill a bowl with rice. Top with the braised pork. Drizzle with some sauce. Serve.

Time-saving method

To braise the pork in a Duromatic, you have to reduce the amount of soy sauce. No water is needed. Turn on the heat and wait until two red bars are visible. Reduce to low heat and cook for 16 minutes. To cook in a Durotherm, you don't need to add water, just cook for 45 minutes. To cook in a Hot pan, add 1/4 cup of water and cook for 25 minutes. Then transfer to outer pot and stew for 30 minutes.

CC 烹調秘訣 | Cooking tips

1. 因每種品牌的醬油鹹度不同，讀者可一邊嘗味道酌量加入。
2. 加入冰糖滷肉才會亮，嘗起來也不會太死甜。
3. 烹調料理用的五香粉可以到中藥房購買。

1. The saltiness of soy sauce may vary, depending on brand. Taste along as you cook the dish.
2. Braising the pork belly with rock sugar not only gives the meat a glossy finish but also prevents it from a dull sweet taste.
3. The five spices powder that is used for cooking can be bought in Chinese herbal medicine stores.

薑母鴨肉糯米飯

Glutinous Rice with Ginger Duck

這是道相當道地的南部風味美食，
QQ的糯米飯散發出濃濃的鴨香味，
總令我想起兒時的回憶。這麼懷舊
的料理，做法絕對不可失傳呀！

This is an authentic Southern Taiwan dish.
The chewy sweet rice with intense aroma
of duck meat always brings me back to the
childhood memories. We must hand down
the recipe of this traditional dish from
generations to generation.

材料

光鴨1隻、老薑120克、長糯米2杯、米酒
11/2杯、麻油3大匙

調味料

雞湯粉或排骨湯粉2大匙

做法

1. 鴨肉剁塊狀；老薑切塊後拍碎。
2. 長糯米洗淨，放入米酒中浸泡約1小時。
3. 將老薑和麻油放入鍋內一起爆香（麻油不可熱），待散發出老薑的香氣再放入鴨肉，炒至鴨肉呈金黃色。
4. 接著倒入長糯米和米酒，放入鴨湯粉拌炒均勻，倒入電鍋內鍋，外鍋倒入2杯水，按下開關煮約30分鐘，再燜約15分鐘即可食用。

這樣做更省時

如果使用休閒鍋或双享鍋，從做法**3.**開始就直接在鍋中料理，放入所有材料後，蓋上鍋蓋，以中小火加熱，待冒煙改小火煮10分鐘後熄火，再移入外鍋燜10分鐘即可，一鍋搞定！

Ingredients

1 whole duck, 120g ginger, 2 cup of long glutinous rice, 11/2 cup of rice cooking wine, 3T sesame oil

Seasonings

2T chicken stock powder or pork-ribs stock powder

Directions

1. Cut the duck meat into pieces. Cut the ginger into pieces and press.
2. Wash the long glutinous rice thoroughly. Soak in rice cooking wine for an hour.
3. Stir-fry the ginger and sesame oil until fragrant (don't heat the sesame oil in advance). Add the duck meat when you smell the aroma of ginger. Stir-fry until golden brown.
4. Next, add the long glutinous rice and the rice cooking wine. Add the duck stock powder and stir evenly. Transfer into the inner pot of a rice cooker. Add 2 cup of water in the outer pot. Press the switch to turn it on and cook for 30 minutes. Allow the rice to rest for 15 minutes before serving.

Time-saving method

When using a Hotpan and Durotherm, you can cook directly in the pan from the direction **3.** After adding all the ingredients, put the lid on and turn on the heat to medium-low heat. When the steam hisses out of the pan, reduce to low heat and cook for 10 minutes. Transfer to the outer pot, simmer for 10 minutes and done. You can have everything done in just one pan!

CC 烹調秘訣 | Cooking tips

1. 麻油不可熱了再放入老薑爆香，否則會苦，所以需用冷鍋、冷油爆香。
2. 麻油料理不可加入鹽調味，否則會有苦味，所以這裡用雞湯粉調味，也可使用排骨湯粉。
3. 除了整隻光鴨（無鴨頭、鴨翅和鴨腳），也可用雞肉來製作。還有一種台灣產的紅面番鴨，更滋補。

1. Never heat the sesame oil before stir-fry ginger until fragrant, otherwise the dish will have a bitter flavor. You need to use cold pot and cold oil to stir-fry ginger.
2. When cooking with sesame oil, never add salt. Otherwise it will taste bitter. That's the reason why we use chicken stock powder for seasonings. Spare-ribs stock powder works well, too.
3. You can substitute chicken for the whole duck (without the head, wings and feet). You can also try replace with Formosa red-face duck, it's even more nourishing.

台式海鮮燴飯
Taiwanese-sytle Rice with Seafood

可隨個人喜好放入海鮮料，所以特別受到朋友和家人的歡迎。只要材料新鮮、實在，怎麼煮都好吃。

This dish is very popular among friends and family members because the seafood ingredients may be changed as desired. As long as the ingredients are fresh and in good quality, it will be tasty no matter how you cook it.

材料

蝦仁30克、透抽300克、胡蘿蔔80克、大白菜200克、黑木耳100克、沙拉筍150克、蝦米2大匙、蔥2支、高湯800c.c.、太白粉1/4杯、香油1大匙、米酒5大匙、白飯適量、香菜適量

調味料

黑醋2 1/2大匙、醬油2大匙、醬油膏2大匙、白胡椒粉少許、鹽少許

做法

1. 蝦仁挑除腸泥；透抽切長寬片狀；胡蘿蔔、大白菜、黑木耳和沙拉筍都切絲；蝦米洗淨，放入2大匙的米酒中浸泡約8分鐘後取出；蔥切段；高湯做法參照p.117。
2. 鍋燒熱，倒入1大匙油，先放入海鮮料拌炒，淋入1大匙的米酒，稍微炒一下後取出。
3. 原鍋再燒熱，倒入2大匙油，先放入蝦米爆香，續入蔥段爆香，放入大白菜，淋入2大匙的米酒煮約1分鐘，倒入高湯煮約2分鐘，再放入胡蘿蔔、黑木耳和沙拉筍煮約3分鐘，倒入調味料和做法**2.**，以太白粉勾芡，淋上香油。
4. 將白飯盛入碗中，淋上做法**3.**，撒上香菜即可食用。

Ingredients

30g shrimps, 300g squid, 80g carrots, 200g Chinese cabbages, 100g black fungus, 150g bamboo shoots, 2T dried shrimp, 2 stalks of green onions, 800c.c. broth, 1/4 cup of tapioca starch, 1T sesame oil, 5T rice cooking wine, rice as needed, coriander as needed

Seasoning

2 1/2T black vinegar, 2T soy sauce, 2T soy sauce paste, white pepper and salt to taste

Directions

1. Remove the sand veins in the shrimps. Cut the squid in wide strips. Shred the carrots, white cabbages, black fungus and bamboo shoots. Wash the dried shrimp then soak in 2T of rice cooking wine for 8 minutes and remove. Cut the green onions in sections. See p.117 for broth recipe.
2. Heat the pan. Add 1T oil. First, stir-fry the seafood ingredients. Pour in 1T of cooking rice. Stir-fry briefly and remove.
3. Heat the same pan. Add 2T oil. Stir-fry the dried shrimp first until fragrant. Add the green onions sections and stir-fry until fragrant. Add the Chinese cabbages and 2T of rice cooking wine and cook for 2 minutes. Next, add the carrots, black fungus and bamboo shoots and cook for 3 minutes. Pour in the seasonings and seafood in direction **2.** Thicken with tapioca starch and drizzle with sesame oil.
4. Fill a bowl with rice. Top with the direction **3.** Sprinkle with coriander and serve.

CC 烹調秘訣 | Cooking tips

1. 沙拉筍在傳統市場和超市都有賣。
2. 黑醋放久會有沉澱物，所以欲使用前需先搖一搖再倒入。

1. Bamboo shoots can be bought at both traditional markets and super markets.
2. Generally, there is sediment in the bottom in black vinegar. Shake it well before use.

什錦菜飯

Combination Fried Rice

一大鍋裡面有菜、有肉多種食材，吃得又飽又營養，省時又方便，推薦給忙碌沒時間做菜的人。如果再準備一鍋湯，全家人一定大滿足。

With various kinds of vegetables and meat all cooked in a pot, this dish is nutritious and can keep you well fed. This dish is simple and easy to make, ideal for people who are too busy to cook. Serving this dish along with a pot of soup will satisfy whole family's appetite.

材料

去骨雞腿2隻、白米2杯（量米杯）、乾香菇8朵、黑木耳100克、洋蔥1/2個、蔥段2支、香菜末適量、高湯2杯、嫩薑3片、紅辣椒1支、蒜末2大匙、蔥絲1支、香油1大匙

醃肉料

醬油2大匙、細砂糖1/3小匙、香菜莖末1大匙、胡椒粉少許、香油1/2大匙

調味料

醬油11/2大匙、白胡椒粉適量

做法

1. 雞腿肉切塊狀，放入醃肉料中拌勻，醃約30分鐘；乾香菇泡水至軟後切絲；黑木耳、洋蔥、嫩薑和紅辣椒都切絲；高湯做法參照p.117。
2. 鍋燒熱，倒入1大匙油，先放入1大匙蒜末爆香，續入白米，以小火炒約2分鐘，倒入高湯拌勻，蓋上鍋蓋，以中小火煮，待冒煙改小火煮約30分鐘。
3. 另一炒鍋燒熱，倒入11/2大匙油，先放入雞肉煎炒至呈金黃色，先取出。再放入洋蔥、辣椒和剩餘的蒜末爆香，加入香菇炒香，放入煎好的雞肉、黑木耳炒勻，加入調味料，最後放入蔥段、嫩薑絲炒勻，再淋上香油。
4. 最後將做法3.鋪在做法2.上面，撒上蔥絲即可食用。

這樣做更省時

做法2.時，如果使用休閒鍋或双享鍋，只要蓋上鍋蓋加熱，待冒煙改小火煮約6分鐘，再移入外鍋燜6分鐘即可。

Ingredients

2 boneless chicken thighs, 2 cup of rice (measurement cup), 8 dried shitake mushrooms, 100g black fungus, 1/2 onion, 2 stalks of green onions sections, minced coriander as needed, 2 cups of broth, 3 sliced tender ginger, 1 red chillies, 2T minced garlic, 1 stalk of shredded green onions, 1T sesame oil

Marinade

2T soy sauce, 1/3 t granulated sugar, 1T minced coriander roots, black pepper to taste, 1/2T sesame oil

Seasonings

11/2T soy sauce, white pepper as needed

Directions

1. Cut the chicken thighs into pieces and marinate for 30 minutes. Soak the dried shitake mushrooms until tender and shred. Shred the black fungus, onions, tender ginger and red chillies. See p.117 for broth recipe.
2. Heat the pan. Add 1T oil. Stir-fry 1T minced garlic until fragrant. Then add the rice and stir-fry on low heat for 2 minutes. Pour in the broth and mix well. Put the lid on and cook on medium-low heat. When the steam hisses out of the pan, reduce to low heat and cook for 30 minutes.
3. Heat another fry-pan. Add 11/2T of oil. Sear the chicken thighs until golden brown and remove. Stir-fry onions, red chillies and the remaining minced garlic until fragrant. Then add the shitake mushrooms and stir-fry until fragrant. Add the sautéed chicken, black fungus and stir evenly ,add seasoning,stalks of green onions and tender ginger and stir evenly, then add sesame oil.
4. Top with the 3. on 2., add green onion.Serve

Time-saving method

When using a Hotpan or Durotherm, after following the direction 2. , just cover and turn on the heat. Reduce to low heat when the steam starts to hiss out of the pan and simmer for 6 minutes. Transfer to outer pot and set aside for 6 minutes. Serve.

 **CC 烹調秘訣** | Cooking tips

這道料理中的薑一定要使用嫩薑味道才會搭。
Make sure to use tender gingers in this dish to bring out the right flavor.

酢醬麵

Noodles with Zha Jiang Sauce

這是我家餐桌上最常出現的麵料
理,我每次炒好一大鍋酢醬料,等
涼後放入玻璃罐中保存,更是朋友
最愛收到的禮物。

This is the most frequently cooked noodle
dish at my home. I always cook a lot of
Zha Jiang Sauce at one time and store in
a glass jar after it cools. It's also one of the
most popular gifts among my friends.

材料

絞肉600克、豆乾200克、蒜末3大匙、蔥末2大匙、沙茶醬11/2大匙、米酒11/2大匙、辣豆瓣醬2大匙、甜麵醬3大匙、香油2大匙、麵條適量、蔥花適量

調味料

醬油1大匙、醬油膏1大匙、老抽2大匙、細冰糖1/2大匙

做法

1. 豆乾切丁。
2. 鍋燒熱，倒入3大匙油，放入豆乾炒至豆乾略帶金黃色，先取出，再放入絞肉炒至均勻地散開，倒入豆乾和沙茶醬拌炒均勻，取出。
3. 原鍋倒入2大匙油，先放入蒜末、蔥末爆香，放入辣豆瓣醬、甜麵醬，炒至散發出香氣，淋入米酒，拌勻後倒入做法2.和調味料，以小火慢炒約15分鐘（途中若太乾，可倒入些許油），淋入香油拌勻成酢醬料。
4. 將麵條煮好，撈出放入碗中，鋪上酢醬料，撒上蔥花，食用時建議淋上少許白醋更美味。

Ingredients

600g ground pork, 200g dried tofu, 3T minced garlic, 2T minced green onions, 11/2T Shacha sauce, 11/2T rice cooking wine, 2T spicy soy bean sauce, 3T sweet bean sauce, 2 T sesame oil, noodles as needed, finely chopped green onions as needed

Seasonings

1T soy sauce, 1T soy sauce paste, 2T dark soy sauce, 1/2T fine rock sugar

Directions

1. Dice the dried tofu into small pieces.
2. Heat the pan. Add 3T oil. Stir-fry the dried tofu until slightly golden brown. Remove. Stir-fry the ground pork until it spreads out evenly. Stir in the dried tofu and Shacha sauce. Remove.
3. In the same pan. Add 2T oil. Stir-fry the minced garlic and minced green onions until fragrant. Add the spicy soy bean sauce and sweet bean sauce. Stir-fry until fragrant. Pour in the rice cooking wine and mix well. Add the meat mixture in direction no.2 and the seasonings. Stir-fry slowly on low heat for 15 minutes (You can add some oil if the sauce is drying out). Drizzle with the sesame oil and the Zha Jiang Sauce is done.
4. Cook noodles and scoop it to a bowl. Top with Zha Jiang Sauce. Sprinkle with the finely chopped green onions. Drizzling with a little bit of white vinegar before serving makes this dish tastes even better.

CC 烹調秘訣 | Cooking tips

1. 老抽是顏色較深的醬油，沒有鹹味，可替食材上色。
2. 甜麵醬和辣豆瓣醬都有鹹度，尤其某些品牌的甜麵醬很鹹，調味時需多斟酌。

1. Dark Soy Sauce is a kind of soy sauce with darker color. It doesn't contain saltiness and is used to add color to a dish.
2. Sweet bean sauce and spicy soy bean sauce both contain saltiness. Some sweet bean sauces are very salty. Be careful not to overadd.

紹子麵

Noodles with Shaozi Sauce

紹子麵是道經典的川味麵料理，四川人把切碎了的食材叫作「紹子」而有此名。醬料的材料豐盛且肉香四溢，滿足了每個饕客的嘴。

This is a classic Sichuan noodle cuisine. People in Sichuan named the minced ingredients "Shaozi". The ingredients in the sauce are substantial and the dish smells so good and meaty. It is something that will satisfy every epicure's appetite.

材料

乾魷魚1/2隻、絞肉600克、黑木耳100克、榨菜200克、紅蔥頭末100克、油蔥酥1/3杯、蔥花1/2杯、煮熟的麵適量

調味料

辣豆瓣醬2大匙、辣椒粉適量、醬油2大匙、白胡椒粉適量、細砂糖1/2小匙、香油2大匙、白醋適量

做法

1. 乾魷魚泡水約30分鐘後切小丁;黑木耳和榨菜都切末。

2. 鍋燒熱,倒入3大匙油,先放入紅蔥頭末炒至呈金黃色,立刻倒入絞肉炒散,加入魷魚炒香,倒入3大匙蔥花稍微炒一下。

3. 接著倒入辣豆瓣醬炒香,放入油蔥酥稍微炒一下,加入辣椒粉、醬油、白胡椒粉、細砂糖炒約6分鐘,再放入黑木耳、榨菜炒約5分鐘,淋上香油炒勻,即成紹子麵醬。

4. 將煮熟的麵放入碗中,鋪上紹子麵醬,淋入白醋,撒上蔥花即可食用。

Ingredients

1/2 dried squid, 600g ground pork, 100g black fungus, 200g pickled mustard, 100g minced shallots, 1/3 cup of deep-fried shallots, 1/2 cup of finely chopped green onions, cooked noodles as needed

Seasonings

2T spicy soy bean sauce, red chillies powder as needed, 2T soy sauce, white pepper as needed, 1/2t granulated sugar, 2T sesame oil, white vinegar as needed

Directions

1. Soak the dried squid in water for 30 minutes. Dice into small pieces. Mince the black fungus and pickled mustard.

2. Heat the pan. Add 3T oil. Stir-fry the minced shallots until golden brown. Immediately stir in the ground pork. Add the squid and stir-fry until fragrant. Add 3T of finely chopped green onions and stir-fry a little bit.

3. Next, stir-fry the spicy soy bean sauce until fragrant. Add the deep-fried shallots and stir-fry a little bit. Add the red chillies powder, soy sauce, white pepper, granulated sugar and stir-fry for 6 minutes. Then add the black fungus and pickled mustard for 5 minutes. Pour in the sesame oil, stir-fry evenly and the "Shaozi" sauce is done.

4. Fill the bowl with cooked noodles. Top with Shaozi sauce. Drizzle with white vinegar. Sprinkle with finely chopped green onions and serve.

CC 烹調秘訣 | Cooking tips

1. 絞肉一定要炒散才會入味,也比較沒有肉腥味。
2. 黑木耳最好是選購乾木耳再浸泡水,口感會比較脆,市售的木耳有時較軟,口感較不佳。
3. 榨菜如果太鹹,使用前需浸泡水約20分鐘,再以清水洗淨,以去除鹹味。

1. Make sure the ground pork does not stick together in a big lump. This will make the meat taste better and remove the odor.
2. When selecting black fungus, dried fungus is a better choice because the texture is crunchier. Sometimes the store-brought fungus might taste mushy, giving it a worse mouthfeel. When using the dried fungus, you need to soak it in water before cooking.
3. If pickled mustards are too salty, soak in the water for 20 minutes before cooking. Then rinse with water to wash away the saltiness.

川味擔擔麵

Sichuan Dan-Dan Noodles

想要在一般小店吃到口感層次分明、正統的擔擔麵已不多見，因為加入了芝麻醬，濃郁的香氣刺激人的食慾。

It's not easy to enjoy a bowl of authentic dan-dan noodles with multiple layers of taste in snack shops nowadays. The key of this dish is the sesame sauce. The strong aroma of the sauce always stimulates people's appetite.

材料

絞肉600克、蝦米末3大匙、蒜末1大匙、榨菜末1/4杯、冬菜末3大匙、辣豆瓣醬2大匙、紹興酒2大匙、香油1大匙、麵條適量、蔥花適量

調味料

醬油2大匙、醬油膏11/2大匙、老抽11/2大匙

拌麵醬

白芝麻醬1大匙、辣油適量、高湯2大匙、醬油1/2大匙、醬油膏1大匙、無糖花生醬1大匙

做法

1. 鍋燒熱，倒入2大匙油，先放入蝦米末爆香，續入蒜末爆香，放入絞肉炒散，再倒入辣豆瓣醬炒香，淋入紹興酒炒約1分鐘，加入榨菜末炒2分鐘，加入調味料炒約5分鐘，最後放入冬菜末炒1分鐘，淋入香油拌勻。
2. 高湯做法參照p.117；將拌麵醬拌勻。
3. 拌麵醬放入容器中，加入煮熟的麵條，再放入適量的做法1.，撒上蔥花即可。

Ingredients

600g ground pork, 3T minced dried shrimp, 1T minced garlic, 1/4 cup of minced pickled mustard, 3T minced pickled cabbages, 2T spicy soy bean sauce, 2T Shaohsing wine, 1T sesame oil, noodles as needed, finely chopped green onions as needed

Seasonings

2T soy sauce, 11/2T soy sauce paste, 11/2T dark soy sauce

Noodles sauce

1T white sesame sauce, chili oil as needed, 2T broth, 1/2T soy sauce, 1T soy sauce paste, 1T unsweetened peanut butter

Directions

1. Heat the pan. Add 2T of oil. First, stir-fry the minced shrimp until fragrant. Then, add and stir-fry the minced garlic until fragrant. Add the ground pork and stir-fry until there are no lumps. Add and stir-fry the spicy soy sauce paste until fragrant. Pour in the Shaohsing wine and stir-fry for 1 minute. Add and stir-fry the minced pickled mustard for 2 minutes. Add and stir-fry the seasonings for 5 minutes. Last, add and stir-fry the minced pickled cabbages for 1 minute. Drizzle with sesame oil and mix well.
2. See p.117 for the broth recipe. Mix the noodle sauce well.
3. Pour the sauce in a mixing bowl. Add the cooked noodles and appropriate amount of the direction 1. Sprinkle with finely chopped green onions and serve.

CC 烹調秘訣 | Cooking tips

1. 做法1.的肉料可以一次先準備多一點的份量，放入冰箱保存，欲食用時只要煮好麵條即可。
2. 不吃辣的人可以選不辣的豆瓣醬，拌麵醬中的辣油直接省略亦可。
3. 無糖花生醬在超市可以買到。

1. You can cook a large portion of mix sauce in direction no.1 at one time and store it in the refrigerator. Next time when you want to enjoy this dish, all you need to do is just to cook the noodles and the meal is ready to serve.
2. For non-spicy eaters: use non-spicy soy bean sauce and skip the chili oil when making the noodle sauce.
3. Unsweetened peanut butter can be found in supermarkets.

台南擔仔麵

Tainan Tan-Tsi Noodles

想吃有名的擔仔麵不必非去台南不可，我要分享我家的私房擔仔麵食譜，絕對不輸正宗擔仔麵，在家也能品嘗府城味喔！

It's not necessary to go to Tainan to savor the famous tan-tsi noodles dish. Here is my secret recipe of tan-tsi noodles. Try it. It tastes as good as authentic one. How wonderful it is to enjoy Tainan cuisine at home!

材料

肉燥料適量、油麵600克、鮮蝦150克、綠豆芽100克、韭菜80克、香菜適量

高湯

豬大骨600克、雞骨600克、黃豆芽200克、洋蔥1個、水2,500c.c.、蝦殼200克

做法

1. 肉燥料做法參照p.101；洋蔥切塊；韭菜切段；豬大骨、雞骨均洗淨。
2. 將高湯的材料放入湯鍋中，熬煮約3小時30分鐘即可。
3. 鮮蝦放入高湯中汆燙熟，取出剝掉蝦殼；綠豆芽汆燙熟；韭菜汆燙一下。
4. 將油麵稍微汆燙一下後取出，瀝乾水分之後放入湯碗中，放上綠豆芽、韭菜，淋入高湯，最後放上肉燥料、鮮蝦，撒上香菜即可食用。

Ingredients

Shitake mushroom flavored pork as needed, 600g oil noodles, 150g shrimps, 100g green bean sprouts, 80g chive, coriander as needed

Broth

600g large pork bones, 600g chicken bones, 200g yellow bean sprouts, 1 onion, 2,500c.c. of water, 200g shrimp shells

Directions

1. See p.101 for the recipe of shitake mushroom flavored pork. Cut the onion in pieces. Cut the chives into sections. Wash the large pork bones and chicken bones.
2. Put all the broth ingredients in a stockpot and stew for 3 hours and 30 minutes.
3. Blanch the shrimps in the broth until cooked thoroughly. Remove and peel the shrimps. Blanch the green bean sprouts until cooked thoroughly. Blanch the chives briefly.
4. Blanch the oil noodles for a little bit, remove, drain and place in a bowl. Add the green sprouts and chives. Pour in the broth. Last, add the shitake mushroom flavored pork, shrimp and sprinkle with the coriander. Serve.

 CC 烹調秘訣 | Cooking tips

1. 做法**2.**中的這種高湯，可以一次熬煮一大鍋，過濾後分裝成數小包，放入冷凍庫中，欲使用時再取出，可用在火鍋湯底、湯麵、粥的高湯。
2. 在擔仔麵中添加貢丸或滷蛋，這碗麵更營養囉！

1. You can cook a big stockpot of broth in direction no.2 at one time and divide it into several bags after straining grease. Store the broth in freezer and take it out before using. You can use it to make hotpot stock, soup noodles and congee.
2. You can add pork balls or braised eggs in tan-tsi noodles to increase the nutrition.

古早風味
魷魚粥

Traditional Congee with Squids

將阿嬤最愛用的菜脯，加入芹菜，烹調而成一碗純樸風味的粥品。一般市面上很難吃得到，只有在家自己做才能回憶古早風味。

This is a simple congee dish cooked with grandma's favorite ingredient, pickled radish, and celery. You don't see this dish that often in restaurants. The only way to savor this traditional dish is to make it by ourselves at home.

材料

乾魷魚200克、五花肉300克、珍珠菜脯280克、白米2杯（量米杯）、紅蔥頭末1/4杯、高湯3,000c.c.、芹菜末1/4杯、細砂糖1/2小匙

調味料

細砂糖1/3小匙、醬油3大匙、白胡椒粉1小匙、鹽適量

做法

1. 乾魷魚洗淨，泡水約30分鐘後取出，切成0.5公分寬、長4公分；五花肉切成寬0.8公分的片狀；高湯做法參照p.117。
2. 鍋燒熱，倒入1大匙油，先放入菜脯炒約3分鐘，續入少許細砂糖炒勻，取出。
3. 原鍋再燒熱，倒入2大匙油，先放入紅蔥頭末爆香至呈金黃色，先取出，立刻放入五花肉煎炒至呈金黃色，加入醬油，撒入白胡椒粉，再倒入魷魚炒香，放入白米炒約1分鐘，待散發出米的香味，倒入高湯煮成粥，倒入調味料拌勻。
4. 放入炒好的菜脯拌勻，盛入碗內，撒上芹菜末即可食用。

這樣做更省時

烹煮粥時，如果使用休閒鍋或双享鍋，放入白米後，蓋上鍋蓋加熱約5分鐘，再移入外鍋燜約10分鐘即可。

Ingredients

200g dried squid, 300g pork belly, 280g dried pearl radish, 2 cups of rice (measurement cup), 1/4 cup of minced shallots, 3,000c.c. broth, 1/4 cup of minced celery, 1/2t granulated sugar

Seasonings

1/3t granulated sugar, 3T soy sauce, 1t white pepper, salt as needed

Directions

1. Wash the dried squid. Soak in water for 30 minutes and remove. Cut into 0.5cm wide and 4cm long strips. Slice the pork belly into 0.8cm wide slices. See p.117 for broth recipe.
2. Heat the pan. Add 1T oil. Stir-fry the dried radish for 3 minutes. Add a pinch of granulated sugar and stir evenly. Remove.
3. Heat the same pan. Add 2T of oil. Stir-fry the minced shallots until golden brown. Remove. Immediately sear the pork belly until golden brown. Add the soy sauce. Sprinkle with white pepper. Add the dried squid and stir-fry until fragrant. Next, stir in the rice and stir-fry for about 1 minute to bring out the aroma of rice. Add the broth and simmer it into congee. Stir in the seasonings.
4. Add the stir-fried dried radish and mix well. Spoon into a bowl. Sprinkle with the minced celery and serve.

Time-saving method

When making the congee in a Hotpan or Durotherm, all you need to do is add the rice, cover and turn on the heat for 5 minutes. Then transfer to out pot and simmer for 10 minutes. Serve.

CC 烹調秘訣 | Cooking tips

1. 菜脯要最後放，口感才會脆。而且，炒菜脯時，放入些許顆粒糖一起炒，口感更佳。
2. 肉的部分建議使用五花肉，利用五花肉的油香味熬煮成的粥會更香。

1. Add the dried radish at the end of the cooking process to keep the crunchy texture. For better texture, stir-fry dried radish along with a little bit of granulated sugar.
2. This dish tastes better with pork belly because the fat of pork belly can bring out a richer flavor.

美味黃金粥

Pumpkin Congee

這是用南瓜烹調的粥品，黃橙橙的顏色、鮮甜的南瓜味，令人食慾大開。

This is a congee made from pumpkins. Its golden color and delicious flavor gives people a good appetite.

材料

南瓜600克、白米2杯（量米杯）、蝦仁200克、生雞胸肉150克、高湯2,500c.c

調味料

鹽適量、白胡椒粉少許

做法

1. 南瓜削除外皮後去籽，切成片狀；蝦仁挑除腸泥後切丁；雞胸肉切末；高湯做法參照p.117。
2. 將白米放入鍋內（不需放油），以小火炒至散發出米香味，放入南瓜、高湯煮至南瓜變成泥糊狀，這時白米已煮成粥了。
3. 接著放入蝦仁、雞胸肉，煮至肉熟，以鹽調味，撒入白胡椒粉即可食用。

Ingredients

600g pumpkin, 2 cups of rice (measuring cup), 200g shrimps, 150g raw chicken breast, 2,500c.c. broth

Seasonings

Salt as needed, white pepper to taste

Directions

1. Peel the pumpkin and remove seeds. Slice the pumpkin. Remove the sand veins of the shrimp. Dice the shrimps. Mince the chicken breast. See p.117 for broth recipe.
2. Add the rice in a pan (don't need to add oil). Stir-fry until fragrant over low heat. Add the pumpkin and broth. Cook until mushed. Now the rice has become congee.
3. Add the shrimps and chicken breast until cooked thoroughly. Add salt and white pepper to taste. Serve.

CC 烹調秘訣 | Cooking tips

1. 烹調料理的過程中，如果有蝦殼、雞骨，可與洋蔥、水熬煮成營養鮮美的高湯，完整運用食材不浪費。
2. 如果要宴客，可加入一些海鮮料，例如花枝丁、鮮魷丁和新鮮干貝，讓這道粥品更豐盛。

1. Don't waste the leftover shrimp shells and chicken bones. You can make a delicious and nutritious broth with these two ingredients along with onions and water.
2. To make this an entertainment dish, you can add some seafood ingredients, such as diced calamari, diced squid and fresh scallops.

豆漿排骨糜米粥

Brown Rice Congee with Soymilk and Pork Ribs

這是一道既香濃又營養的健康養生粥品，很少人能抵擋它的美味，是中式早餐最好的選擇。

This is the best choice of Chinese breakfast. It's nutritious, contains rich flavor and wins almost everyone's heart.

材料

小排骨600克、糜米250克、無糖豆漿800c.c.、高湯1,000c.c.、去皮牛蕃茄3個、胡蘿蔔1條

調味料

鹽適量

做法

1. 小排骨倒入滾水中汆燙一下，取出洗淨。也可將小排骨移入烤箱中，以上下火230℃烤至稍微呈金黃色。
2. 牛蕃茄切塊；胡蘿蔔切滾刀塊；高湯做法參照p.117。
3. 將小排骨倒入鍋內，放入糜米、牛蕃茄和胡蘿蔔，倒入高湯和豆漿，以大火煮滾，再改小火煮約40分鐘，以鹽調味即可食用。

Ingredients

600g baby back ribs, 250g brown rice, 800c.c. unsweetened soymilk, 1,000c.c. broth, 3 peeled beef tomatoes, 1 carrot

Seasonings

Salt as needed

Directions

1. Blanch the baby back ribs in boiling water for a little while. Remove and rinse thoroughly. The alternative method is to bake the ribs in the oven at 230°C until slightly golden brown.
2. Cut the beef tomatoes. Roll cut the carrot. See P.117 for broth recipe.
3. Place the ribs in a pot. Add the brown rice, beef tomatoes and carrots. Pour in the broth and soymilk. Bring to a boil on high heat and reduce to low heat for 40 minutes. Add salt to taste and serve.

✤ CC 烹調秘訣 | Cooking tips

1. 無糖豆漿又叫清漿，在一般超市或豆漿店有售。
2. 多加入一些小魚乾去煮，可增加鈣質的攝取。

1. Unsweetened soymilk is also known as "tsing-chiang" in Mandarin and can be bought at supermarkets or soymilk shops.
2. Cooking with some dried fingerling can increase the consumption of calcium.

Soup Section

Part 5 湯類

一頓餐裡，如果少了湯，總覺得有點美中不足；
飯後喝碗湯，這一餐才能畫下完美的句點。
尤其天涼的時候，
一碗湯下肚，真是暖到心坎裡。
湯，也許不是餐桌上的主角，
但絕對是不可或缺的要角，
甚至，沒食慾的時候，來碗湯就很滿足。

A meal without soup is like a cloud on one's happiness.
Having a bowl of soup at the end of meal gives you a feeling of perfect ending.
It's a warm treat especially in cold days.
Soup may not play a starring role on dining tables
but is definitely an important part.
Next time when you don't feel much of appetite,
you might find a feeling of satisfaction by just having a bowl of soup.

莧菜魩仔魚羹

Amaranth and Miniature Silver Fish in Thick Soup

這道鮮魚羹是我以前每次到鼻頭角或蘇澳遊玩時必點的一道菜，為了讓大家也能品嘗到美味，建議在家自己動手做吧！同時還兼具衛生健康。

In the past, every time I visited Bitou Cape or Su-ao Town, I would order this fish soup. Let's make this dish at home so that everyone can not only enjoy the tasty soup but also eat healthy.

材料

莧菜300克、魩仔魚100克、蒜末1大匙、芹菜末1大匙、高湯1,000c.c.、太白粉3大匙

調味料

鹽適量、胡椒粉少許

做法

1. 莧菜切末；高湯做法參照p.117。
2. 鍋燒熱，倒入1大匙油，先放入蒜末爆香，續入高湯，加入莧菜煮至變軟，再加入魩仔魚煮一下。
3. 放入調味料拌勻，加入太白粉勾芡，盛入碗中，撒上芹菜末即可食用。

Ingredients

300g amaranth, 100g miniature silver fish, 1T minced garlic, 1T minced celery, 1,000c.c. broth, 3T tapioca starch

Seasonings

Salt as needed, pepper to taste

Directions

1. Mince the amaranth. See p. 117 for the broth recipe.
2. Heat the pan. Add 1T oil. Stir-fry the minced garlic until fragrant first. Then add the broth and amaranth. Cook until tender. Add the miniature silver fish and cook briefly.
3. Add the seasoning and mix well. Thicken with the tapioca starch. Spoon into a bowl and sprinkle with the minced celery. Serve.

CC 烹調秘訣 | Cooking tips

莧菜需煮得軟透才會釋放出香味。此外，這道料理口感滑順，適合家中有長輩和小孩的家庭。

Amaranth needs to be cooked until tender to release the flavor. With its silken texture, this soup is especially good for families with seniors and youngsters.

山藥蛤蜊雞湯

Chicken Soup with Yam and Clams

這是一道很適合在露營野餐時炊煮的湯品，做法非常簡單，尤其營地附近若有販售土雞，以土雞來烹調，湯汁更鮮甜，當然宴客也會更受歡迎。

This is an ideal soup at camping. The directions are very simple. If you happen to see native chicken vendors near campsite, go for it. The soup tastes more delicious when cooking with a native chicken. This is also a very popular entertaining dish.

材料

土雞1隻（約1.8公斤、3台斤）、山藥300克、蛤蜊600克、嫩薑5片、紅棗8粒、枸杞1大匙、水2,000c.c.

調味料

鹽適量

做法

1. 土雞剁成塊；山藥削除外皮後切成滾刀塊。
2. 將土雞放入鍋中，倒入水，放入薑片煮約30分鐘，續放入山藥，煮約3分鐘，再放入蛤蜊、紅棗和枸杞，煮至蛤蜊殼開就熄火，以鹽調味即可食用。

Ingredients
1 native chicken (about 1.8kg), 300g yam, 600g clams, 5 sliced tender ginger, 8 red dates, 1T goji berries, 2,000c.c. water

Seasonings
Salt as needed

Directions
1. Chop the native chicken into pieces. Peel and cut the yam into wedges.
2. Place the chicken into a stockpot. Add water and sliced ginger. Cook for 30 minutes. Then add the yam and cook for 3 minutes. Next, add the clams, red dates and goji berries until clams open their shells. Turn off the heat. Add salt to taste and serve.

CC 烹調秘訣 | Cooking tips

1. 製作這道料理時，一般人汆燙雞的原因是為了去掉雜質，但我認為這樣更容易導致雞天然的甜味流失掉，非常可惜。所以我建議不要汆燙，只要在烹調過程中用細網勺撈除雜質和油質即可。這樣就不需特別煮一鍋滾水，還可以省下不少時間。
2. 這道料理中的薑需使用嫩薑才對味，老薑一點都不合。

1. People usually blanch the chicken in boiling water to get rid of impurities. But in my opinion, blanching the chicken tends to lose its natural tasty flavor easily. Therefore, I don't recommend people to blanch the meat. Just use a skimmer to skim the impurities and fats during the cooking process. You can not only save the hassle of boiling a pot of water but also save a lot of time.
2. The ginger used in this dish must be tender ginger. Old ginger doesn't go well with it.

瓜仔雞湯

Chicken Soup with Pickled Cucumber

帶點懷舊風味、鮮甜的口感，是許多居住在國外的人最懷念台灣的家鄉菜。

This is a traditional dish with delicious flavor. It's is one of Taiwanese cuisine that people living overseas miss the most.

材料

土雞1/2隻、玻璃罐裝花瓜含汁1/2瓶、醃蔭瓜或醃冬瓜3大匙、嫩薑3片、蔥2支、高湯或水2,000c.c.

調味料

鹽適量

做法

1. 土雞剁塊狀；蔥切段；高湯做法參照p.117。
2. 將高湯倒入鍋中，放入土雞、薑片和蔥段煮滾，過程中不時撈除雜質和油質，以小火煮約30分鐘。
3. 將花瓜和醃蔭瓜倒入土雞鍋中，再煮約5分鐘，以鹽調味即可食用。

Ingredients

1/2 native chicken, 1/2 jar pickled cucumber with juice, 3T pickled melon or pickled winter gourd, 3 sliced tender ginger, 2 stalks of green onions, 2,000c.c. broth or water

Seasonings

Salt as needed

Directions

1. Chop the native chicken into pieces. Cut the green onions into sections. See p.117 for the broth recipe.
2. Pour the broth into a stockpot. Add the chicken, sliced ginger and green onions sections and bring to a boil. Skim impurities and fat that rise to the top. Cook for 30 minutes over low heat.
3. Add the pickled cucumber and pickled melon into the pot. Cook for another 5 minutes. Add salt to taste and serve.

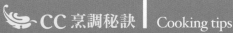

CC 烹調秘訣 | Cooking tips

1. 花瓜汁和醃冬瓜有鹹度，調味食鹽的用量要特別注意。
2. 也可以用杏鮑菇片代替花瓜，口味清淡一樣美味。

1. The pickled cucumber juice and pickled winter gourd are already salty. Therefore, don't overadd salt.
2. You can substitute king oyster mushroom over pickled cucumber. The soup will still taste light and delicious.

胡椒豬肚湯

Pig Stomach Soup with White Peppercorns

這道簡單、好喝又有益腸胃的湯品，是我家的私房菜，僅需簡單幾種材料，就能烹調出不輸專業廚師的口味。

This is my home dish. It is delicious, easy to make and good for the digestive system. With just a few ingredients, you can make a soup that tastes as good as the professional chef's.

材料

豬肚1副、白胡椒粒4大匙、米酒11/2大匙、麵粉3大匙、鹽1大匙、油11/2大匙、水或高湯2,000c.c.、小白布袋1個

調味料

鹽少許

做法

1. 豬肚用麵粉、鹽和油稍微搓洗一下，再以清水洗淨，放入滾水中汆燙一下後撈出。
2. 將白胡椒粒放入小布袋內，綁緊，放入豬肚內。
3. 將水或高湯倒入鍋中，放入做法2.煮約1小時20分鐘，取出切成條狀，再放回湯內，以鹽調味，淋入米酒即可食用。

Ingredients
1 pig stomach, 4T white peppercorns, 11/2T rice cooking wine, 3T flour, 1T salt, 11/2T oil, 2,000c.c. water or broth, 1 white cloth bag

Seasonings
Salt to taste

Directions
1. Briefly rub the pig stomach with flour, salt and oil. Rinse well in running water and blanch in boiling water. Remove.
2. Put the white peppercorns in a piece of white cloth bag. Tie tightly and stuff in the pig stomach.
3. Pour water or broth in a stockpot. Add the pig stomach in direction 2. and cook for 1 hour and 20 minutes. Remove and cut into strips. Place back in the soup. Season with salt and pour on the rice cooking wine.

 CC 烹調秘訣 | Cooking tips

白胡椒粒、小白布袋在中藥房都買得到。

White peppercorns and white cloth bags can be found in Chinese herbal medicine store.

香菇肉羹
Shiitake Mushroom and Pork in Thick Soup

小吃店的香菇肉羹食材、做法不同，各種風味兼具。我家的做法是較具古早風味，而且料多實在，是餐桌上不可缺的一道料理。

The ingredients and cooking method of shitake mushrooms and pork in thick soup are different in every snack shop with its own signature flavor. The method that I use is more traditional, plus a lot of good ingredients, thus this soup is a must-have dish on our dining table.

材料

小里肌300克、魚漿500克、乾香菇8朵、竹筍150克、胡蘿蔔100克、高湯或水2,000c.c.、香油2大匙、太白粉1/4杯、蒜泥3大匙、紅蔥頭酥（油蔥酥）1/4杯、香菜適量

醃肉料

醬油2大匙、香油1大匙、細砂糖1/3小匙、五香粉1/3小匙、米酒1大匙、白胡椒粉少許、太白粉4大匙、冷高湯或水4大匙

調味料

醬油2大匙、醬油膏1大匙、白砂糖1/3大匙、鹽適量、黑醋2大匙

做法

1. 小里肌切約長3公分、寬0.7公分，放入醃肉料中醃約30分鐘，然後和魚漿拌勻。
2. 乾香菇泡水至軟後切絲；竹筍和胡蘿蔔都切絲；高湯做法參照p.117；紅蔥頭酥的做法參照p.99。
3. 每一塊小里肌都裹上魚漿，再一塊塊放入高湯中煮至肉羹浮起，表示肉羹熟了。
4. 將香菇、竹筍和胡蘿蔔放入高湯內煮約4分鐘，放入調味料拌勻，加入太白粉勾芡，淋上香油。
5. 將香菇肉羹盛入碗內，鋪上適量的蒜泥、紅蔥頭酥，撒上香菜即可食用。

Ingredients

300g tenderloin, 500g fish paste, 8 dried shiitake mushrooms, 150g bamboo shoots, 100g carrots, 2,000c.c. broth or water, 2T sesame oil, 1/4 cup tapioca starch, 3T grinded garlic, 1/4 cup deep-fried shallots, coriander as needed

Marinade

2T soy sauce, 1T sesame oil, 1/3 t granulated sugar, 1/3 t five spices powder, 1T rice cooking wine, white pepper to taste, 4T tapioca starch, 4T cold broth or water

Seasonings

2T soy sauce, 1T soy sauce paste, 1/3T white granulated sugar, salt as needed, 2T black vinegar

Directions

1. Cut the tenderloin into 3-cm long, 0.7-cm wide strips. Marinate for 30 minutes then mix well with fish paste.
2. Soak the dried shiitake mushrooms until tender and shred. Shred the bamboo shoots and carrots. See p.117 for the broth recipe. See p.99 for the deep-fried shallot recipe.
3. Coat each pork strips with fish paste. Place into broth one by one and cook until the meat floats up, indicating that it's cooked thoroughly.
4. Cook the shiitake mushrooms, bamboo shoots and carrots in broth for 4 minutes. Add the seasoning and mix well. Thicken with tapioca starch. Drizzle with sesame oil.
5. Spoon the thick soup into a serving bowl. Add the grinded garlic and deep-fried shallots as need. Sprinkle with coriander and serve.

CC烹調秘訣 | Cooking tips

1. 小里肌的口感較後腿肉來得軟嫩，可依個人喜好選擇。
2. 加入燙熟的油麵，就成了一碗香噴噴的香菇肉羹麵囉！
3. 可以一次多製作一些肉羹，室溫放涼後再放入冷凍保存，食用前再取出即可。

1. The texture of pork tenderloin is more tender pork legs. Choose as desired.
2. For variation, you may serve the soup with cooked oil noodles.
3. You can cook a certain amount of thick soup at one time. Store in the freezer after cooling down and thaw before serving.

酸辣湯

Hot and Sour Soup

記得我剛學習烹調中菜時，一位名廚告訴我美味的酸辣湯，一定少不了黑白醋和黑白粉，也就是黑醋、白醋、黑胡椒粉和白胡椒粉。加入這些調味料，我的酸辣湯變成所向無敵了。

I remember when I started to learn cooking Chinese cuisine, a famous chef told me that a bowl of delicious hot and sour soup must have black/white vinegar and black/white powder. That means black vinegar, white vinegar, black pepper and white pepper. After adding these seasonings, my hot and sour soup turned into the most delicious one that I had ever tasted.

材料

肉絲150克、豆腐1塊、黑木耳80克、胡蘿蔔80克、竹筍80克、雞蛋2個、太白粉1/2杯、高湯1,800c.c.、香油2大匙、蔥花適量、香菜適量

醃肉料

醬油1大匙、米酒1/2大匙、香油1/3大匙、白胡椒粉少許

調味料

醬油2大匙、醬油膏2大匙、鹽適量、白醋11/2大匙、黑醋2大匙、白胡椒粉適量、黑胡椒粉1/2大匙

做法

1. 將肉絲放入醃肉料中醃約20分鐘。
2. 豆腐、黑木耳、胡蘿蔔和竹筍都切絲；雞蛋打散；高湯做法參照p.117。
3. 高湯倒入湯鍋中煮滾，放入豆腐、黑木耳、胡蘿蔔和竹筍煮約3分鐘，再加入肉絲煮約2分鐘，加入調味料煮滾。
4. 將湯杓放入湯中間，以畫圓的方式攪動，一邊不停地攪動，一邊慢慢倒入蛋液，此時會呈蛋花狀，再以畫圓的方式倒入太白粉勾芡，一邊慢慢攪動，勾芡才不會凝結成塊。
5. 撒上香油、蔥花和香菜即可食用。

Ingredients

150g pork strips, 1 square of tofu, 80g black fungus, 80g carrots, 80g bamboo shoots, 2 eggs, 1/2 cup tapioca starch, 1,800c.c. broth, 2T sesame oil, finely chopped green onions as needed, coriander as needed

Marinade

1T soy sauce, 1/2T rice cooking wine, 1/3 T sesame oil, white pepper to taste

Seasonings

2T soy sauce, 2T soy sauce paste, salt as needed, 11/2T white vinegar, 2T black vinegar, white pepper as needed, 1/2T black pepper

Directions

1. Marinate the pork strips for 20 minutes.
2. Shred the tofu, black fungus, carrots and bamboo shoots. Whisk the eggs. See p.117 for the broth recipe.
3. Pour the broth in a stockpot and bring to a boil. Add the tofu, black fungus, carrots and bamboo shoots for 3 minutes. Then add the pork strips and cook for 2 minutes. Add the seasonings and bring to a boil.
4. Place the ladle in the middle of soup and stir in circle. As you stir constantly, slowly drizzle in the whisked egg. Once the eggs have been dropped, thicken the soup with tapioca starch in circle. Stir gradually to avoid lumps.
5. Sprinkle with finely chopped green onions and corianders. Drizzle with sesame oil and serve.

CC 烹調秘訣 | Cooking tips

1. 勾芡的要訣是要用中大火，湯勺需不停地攪動，否則太白粉會結塊。
2. 非竹筍的產季建議可加些金針菇，湯料更豐富。

1. When thickening the soup, you must do it over medium-high heat and continually stir the ladle to avoid tapioca starch lumps.
2. You can add some enoki mushrooms if bamboo shoots are not in season.

清燉牛肉湯

Stewed Beef Soup

常有學生向我抱怨：「美味的清燉
牛肉湯實在太難做了！」其實不需
特別加入中藥，只要以洋蔥的甜、
八角的香來提味，自然湯鮮味美。

Students often complain to me, "it's too
difficult to make delicious stewed beef
soup!" In fact, you don't have to add
Chinese herbs, just use onions and anises
to bring out the flavor then you will have
a bowl of good and delicious soup.

材料

牛腱1,000克、洋蔥2個、嫩薑1塊、月桂葉1片、八角2粒、薑片5片、蔥花適量、牛高湯或水2,500c.c.。

牛高湯

牛骨1,200克、水3,000c.c.、嫩薑片2片、洋蔥1個、蔥3支

做法

1. 牛腱切成厚1～1.2公分；洋蔥切塊狀；嫩薑切絲。
2. 先製作牛高湯：將洗淨的牛骨、水、嫩薑片、洋蔥塊和蔥段放入鍋中熬煮約5小時即可。如果使用壓力鍋，更能縮短時間，1小時就OK。
3. 將水或牛高湯倒入鍋中，放入洋蔥、月桂葉、八角和薑片，再倒入牛腱，煮滾後改小火煮約2小時。
4. 煮好的牛肉湯盛入湯碗中，撒上蔥花、薑絲即可食用。也可加入麵或冬粉，味道很讚喔！

~~~

這樣做更省時

如果使用壓力鍋燉煮，加熱後等上升兩條紅線，再改小火煮約23分鐘即可。

Ingredients

1,000g beef shank, 2 onions, 1 piece of tender ginger, 1 bay leaf, 2 anises, 5 slices of ginger, finely chopped green onions as needed, 2,500c.c. beef broth or water

Beef broth

1,200g beef bone, 3,000c.c. water, 2 slices of tender ginger, 1 onion, 3 stalks of green onions
Place all the ingredients in a stockpot and stew for 5 hours.

Directions

1. Cut the beef shank into 1 to 1.2-cm thick pieces. Cut the onions into pieces. Shred the tender ginger.
2. First make the beef broth: Place the rinsed beef bone, water, sliced tender ginger, onion pieces and green onion sections into a stockpot and stew for 5 hours. Using a Duromatic can shorten the cooking time to 1 hour.
3. Pour water or beef broth into a stockpot. Add the onions, bay leaf, anises, sliced ginger and beef shank. Bring to a boil then reduce to low heat for 2 hours.
4. Spoon the cooked beef soup into a bowl. Sprinkle with finely chopped green onions and shredded ginger and serve. The soup goes very well with noodles and green bean noodles.

Time-saving method

To use a presser cooker, just turn on the heat then reduce to low heat for 23 minutes when two red bars are visible.

CC 烹調秘訣 | Cooking tips

1. 一般中式料理會將牛肉放入滾水中汆燙，是為了去雜質，但我認為這個動作會導致牛肉的鮮美味流失在汆燙的水中，非常可惜。除了一般的汆燙法，我建議以下兩個方法：一是在燉煮時，用網勺一邊撈出雜質和油質；二則是將牛肉放入烤箱，以上下火220℃烤至稍微金黃即可，這樣不僅在燉煮時不會有雜質，還可使肉更香。
2. 薑的外皮常被丟棄，建議可將嫩薑外皮較粗的部分切成片，放入清燉牛肉湯中一起煮，軟嫩的部分則切絲使用，避免產生太多的廚餘。

1. In Chinese cooking, we usually blanch the beef in boiling water to get rid of impurities. But in my opinion, blanching the beef in boiling water results in losing the delicious flavor. It's such a waste. In addition to the common blanching method, here are two other ways that I recommend: one is to use a skimmer to skim impurities and fat during stewing. The other one is to bake the beef in an oven at 220°C until slightly golden brown. These two are good ways not only to skim impurities but also enhance the flavor.
2. We usually discard ginger skins right away. In order to minimize the food waste, we can slice the rougher part of ginger skins and cook with the beef soup. Then shred the tender part of it for further use.

KUHN RIKON
SWITZERLAND

SWISS made
by KUHN RIKON

瑞康國際企業股份有限公司
Tel 886 2 2810 8580
Fax886 2 8811 2518

港澳區總代理
大象行批發有限公司
Tel 852 2543 4296
Fax852 2543 1128

北區
士林旗艦店1F
Tel 886 2 2885 0038

太平洋SOGO百貨中壢元化店7F
Tel 886 3 427 0723

中區
新光三越百貨台中中港店8F
Tel 886 4 2252 1861

銅鑼灣
崇光百貨7F家品部
Tel 852 2831 8634

生活館
灣仔店
Tel 852 2872 8738

太平洋SOGO百貨台北忠孝館8F
Tel 886 2 8771 3442

太平洋SOGO百貨新竹店9F
Tel 886 3 522 6834

HOLA和樂家居館台中中港店2F
Tel 0985 713 263

中區
先施百貨3F家品部
Tel 852 2545 8366

銅鑼灣店
Tel 852 2506 0001

太平洋SOGO百貨台北復興館8F
Tel 886 2 8772 1285

BIG CITY遠東巨城購物中心6F
Tel

FE'21台中大遠百
Tel

荃灣
千色店3F家品部
Tel 852 2413 8686

沙田店
Tel 852 2687 0333

新光三越信義新天地A8館7F
Tel 886 2 2722 7178

HOLA和樂家居館台北內湖店1F
Tel 0985 713 260

南區
新光三越百貨台南西門店B1
Tel 886 6 303 0393

先施百貨GF家品部
Tel 852 2258 3688

尖沙咀店
Tel 852 2314 8821

新光三越台北南西店7F
Tel 886 2 2567 8316

HOLA和樂家居館士林店B1
Tel 0985 713 257

新光三越百貨高雄左營店9F
Tel 886 7 345 4486

旺角
先施百貨4F家品部
Tel 852 2399 0688

澳門
新八佰伴6F家品部
Tel 853 2872 5338

桃園FE'21遠東百貨新概念店10F
Tel 886 3 332 6923

HOLA和樂家居館桃園南崁店1F
Tel 886 3 332 9205

HOLA和樂家居館台南仁德店2F
Tel 0985 713 262

板橋FE'21遠東百貨新概念店
Tel

HOLA和樂家居館高雄左營店1F
Tel 0985 713 262